Lecture Notes
in Business Information Processing   34

W0230437

Antonia Albani   Joseph Barjis
Jan L.G. Dietz (Eds.)

Advances in
Enterprise Engineering III

5th International Workshop, CIAO! 2009, and
5th International Workshop, EOMAS 2009, held at CAiSE 2009
Amsterdam, The Netherlands, June 8-9, 2009
Proceedings

 Springer

Volume Editors

Antonia Albani
Jan L.G. Dietz
Delft University of Technology
Mekelweg 4, 2628 CD Delft, The Netherlands
E-mail: {a.albani,j.l.g.dietz}@tudelft.nl

Joseph Barjis
Delft Unitersity of Technology
Jaffalaan 5, 2628 BX Delft, The Netherlands
E-mail: J.Barjis@TUDelft.NL

Library of Congress Control Number: Applied for

ACM Computing Classification (1998): J.1, H.4, H.1, D.2, I.6

ISSN   1865-1348
ISBN-10   3-642-01914-5 Springer Berlin Heidelberg New York
ISBN-13   978-3-642-01914-2 Springer Berlin Heidelberg New York

springer.com

© Springer-Verlag Berlin Heidelberg 2009
Printed in Germany

Typesetting: Camera-ready by author, data conversion by Scientific Publishing Services, Chennai, India
Printed on acid-free paper   SPIN: 12685660   06/3180   5 4 3 2 1 0

Preface

In the era of continuous changes in internal organizational settings and external business environments – such as new regulations and business opportunities – modern enterprises are subject to extensive research and study.

For the understanding, design, and engineering of modern enterprises and their complex business processes, the discipline of enterprise engineering requires sound engineering principles and systematic approaches based on rigorous theories. Along with that, a paradigm shift seems to be needed for addressing these issues adequately. The main paradigm shift is the consideration of an enterprise and its business processes as a social system. In its social setting, an enterprise and its business processes represent actors with certain authorities and assigned roles, who assume certain responsibilities in order to provide a service to its environment. Second to that, a paradigm shift is to look at an enterprise as an artifact purposefully designed for a certain mission and goal.

The need for this paradigm shift, along with the complexity and agility of modern enterprises, gives inspiration for the emerging discipline of *enterprise engineering* that requires development of new theories and methodologies. To this end, the prominent methods and tools of *modeling* and *simulation* play a significant role. Both (conceptual) modeling and simulation are widely used for understanding, analyzing, and engineering an enterprise (its organization and business processes).

In addressing the current challenges and laying down some principles for enterprise engineering, this book, following the success of its first volume in 2008, includes a collection of papers presented and discussed at the co-located meeting of CIAO! 2009 and EOMAS 2009, organized in conjunction with the CAiSE 2009 conference. The scopes of these two workshops are to a large extent complementary, with CIAO! being more focused on the theory and application of enterprise engineering and EOMAS on the methods and tools for modeling and simulation.

June 2009

Antonia Albani
Joseph Barjis
Jan L.G. Dietz

An Introduction to Enterprise Engineering

The Paradigm Shift

Enterprise engineering is an emerging discipline that studies enterprises from an engineering perspective. The first paradigm of this discipline is that enterprises are purposefully designed and implemented systems. Consequently, they can be re-designed and re-implemented, if there is a need for change. All kinds of changes are accommodated: strategic, tactical, operational, and technological. The second paradigm of enterprise engineering is that enterprises are social systems. This means that the system elements are social individuals, and that the essence of an enterprise's operation lies in the entering into and complying with commitments between these social individuals[1].

The Theoretical Roots

Enterprise engineering is rooted in both the organizational sciences and the information system sciences. Three concepts are already paramount to the theoretical and practical pursuit of enterprise engineering: enterprise ontology, enterprise architecture, and enterprise governance. *Enterprise ontology* concerns the understanding of an enterprise in a way that is fully independent of any implementation. The (one and only) ontological model of an enterprise shows the essence of its operation. It is the starting point for designing and implementing all kinds of changes. It is also extremely stable over time; most changes appear to be changes in the implementation. *Enterprise architecture* concerns the identification, the specification, and the application of design principles, which come in addition to the specific requirements in every change project. Design principles are the operational shape of an enterprise's strategic basis (mission, vision). Only in this way can one achieve and guarantee that the operations of an enterprise are fully compliant with its mission and strategies. Lastly, *enterprise governance* constitutes the organizational conditions for incorporating enterprise ontology and enterprise architecture in an enterprise's practice. It constitutes the primary condition for making the enterprise engineering approach feasible and beneficial.

The Current Evidence

The vast majority of strategic initiatives fail, meaning that enterprises are unable to gain success from their strategy. The high failure rates are reported from

---

[1] Basically and principally, only humans can take the role of social individual. We do recognize, however, the increasing belief among researchers that in the future artifacts could also take this role.

various domains: total quality management, business process reengineering, six sigma, lean production, e-business, customer relationship management, as well as from mergers and acquisitions. It appears that these failures are mostly the avoidable result of an inadequate implementation of the strategy. Rarely are they the inevitable consequence of a poor strategy. Abundant research indicates that the key reason for strategic failures is the lack of coherence and consistency, collectively also called congruence, among the various components of an enterprise. At the same time, the need to operate as an integrated whole is becoming increasingly important. Globalization, the removal of trade barriers, deregulation, etc., have led to networks of cooperating enterprises on a large scale, enabled by the virtually unlimited possibilities of modern information and communication technology. Future enterprises will therefore have to operate in an ever more dynamic and global environment. They need to be more agile, more adaptive, and more transparent. In addition, they will be held more publicly accountable for every effect they produce. These challenges are traditionally addressed with black-box thinking-based knowledge, i.e., knowledge concerning the function and the behavior of enterprises, as contained in the organizational sciences. Such knowledge is sufficient, and perfectly adequate, for managing an enterprise (within the range of control). However, it is definitely inadequate for changing an enterprise. In order to bring about changes, white-box-based knowledge is needed, i.e., knowledge concerning the construction and the operation of enterprises. Developing and applying such knowledge requires no less than a paradigm shift in our thinking about enterprises, since the organizational sciences are dominantly oriented towards organizational behavior, based on black-box thinking.

The Evolutionary Milestones

The current situation in the organizational sciences resembles very much the one that existed in the information system sciences around 1970. At that time, a revolution took place in the way people conceived information technology and its applications. Since then, people have been aware of the distinction between the *form* and the *content* of information. This revolution marks the transition from the era of data systems engineering to the era of information systems engineering. The comparison we draw with the information system sciences is not an arbitrary one. On the one hand, the key enabling technology for shaping future enterprises is the modern information and communication technology (ICT). On the other hand, there is a growing insight into the information systems sciences that the central notion for understanding profoundly the relationship between organization and ICT is the entering into and complying with commitments between social individuals. These commitments are raised in communication, through the so-called *intention* of communicative acts. Examples of intentions are requesting, promising, stating, and accepting. Therefore, as the content of communication was put on top of its form in the 1970s, the intention of communication is now put on top of its content. It explains and clarifies the organizational notions of collaboration and cooperation, as well as authority and responsibility. It also puts organizations definitely in the category of social systems, very distinct from information systems. Said revolution in the information systems sciences marks the transition from the era of information systems engineering to the era of enterprise engineering, while at the same time merging with relevant parts of the organizational sciences, as illustrated in Fig. 1.

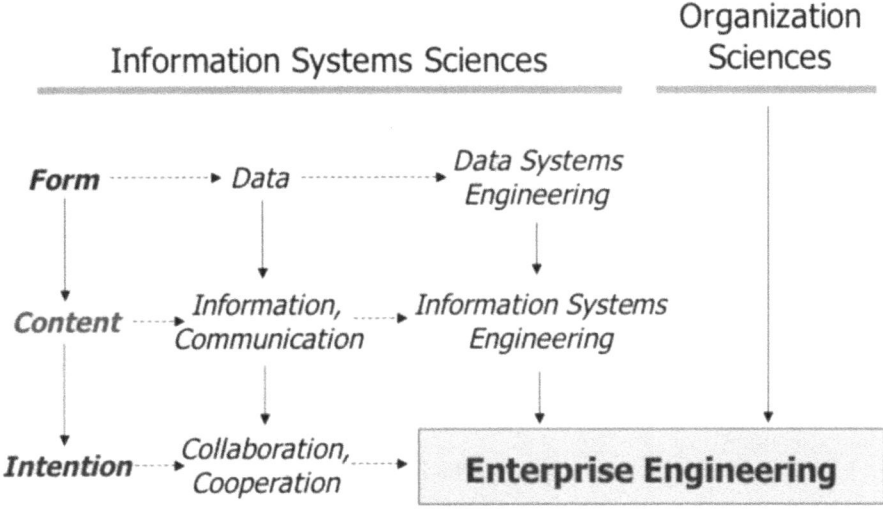

Fig. 1. Enterprise engineering

The mission of the discipline of enterprise engineering is to combine (relevant parts from) the organizational sciences and the information system sciences, and to develop theories and methodologies for the analysis, design, and implementation of future enterprises. Two crucial concepts have already emerged that are considered paramount for accomplishing this mission: enterprise ontology and enterprise architecture. A precondition for incorporating these methodologies effectively in an enterprise is the good establishment of enterprise governance.

Theoretically, *enterprise ontology* is the understanding of an enterprise's construction and operation in a fully implementation-independent way. Practically, it is the highest-level constructional model of an enterprise, the implementation model being the lowest one[2]. Compared to its implementation model, the ontological model offers a reduction of complexity of well over 90%. Only by applying this notion of enterprise ontology can substantial changes of enterprises be made intellectually manageable.

Theoretically, *enterprise architecture* is the normative restriction of design freedom. Practically, it is a coherent and consistent set of principles that guide the (re)design and (re)implementation of an enterprise, and that come in

---

[2] Dietz, J.L.G., Enterprise Ontology – Theory and Methodology, Springer, 2006, ISBN 978-3-540-29169-5

addition to the specific requirements in a change project[3]. These principles are derived from the enterprise's strategic basis (mission, vision). Only by applying this notion of enterprise architecture can consistency be achieved between the strategic basis and the operational business rules of an enterprise.

Enterprise governance is the organizational competence for continuously exercising guiding authority over enterprise strategy and architecture development, and the subsequent design, implementation, and operation of the enterprise[4]. Adopting this notion of enterprise governance enables an enterprise to be compliant with external and internal rules, and to perform in an optimal and societally responsible way.

Modeling and Simulation

Every time that a change happens in the business environment or a change is required due to certain circumstances, it results in analysis and design of some aspects of the enterprise (organization, business processes, supporting technology, etc.). Current trends in business process management show that processes-oriented approaches are receiving increasing attention in analyzing and designing enterprises and implementing innovations addressing the external forces (customers, competitors, environment, etc.). As the very core of process innovation is change, and changes always need to be evaluated in comparison with different scenarios and situations, this demands an even more integral role of modeling and simulation in design, redesign, and process improvement activities of enterprise engineering. Obviously any change is risky and may have serious consequences for enterprises. Early mitigation of risks associated with redesign and innovation is highly important, especially in situations with many uncertainties. Here is where modeling and simulation play an enormous role in the analysis, design, redesign, comparison of alternatives, and measurement of the effects of changes[5].

Ontology-Based Development of Information Systems

Based on the notion of enterprise engineering, new modeling methodologies are needed to cope with the specific aspects of an enterprise as a designed and engineered artifact. Such methodologies should not only comprise methods and models to design the enterprise in order to understand and change it, but also to design and implement information systems supporting the operations and decision makings of such enterprises. Several enterprise modeling methodologies exist and are widely applied in practice today. But most of them are not

---

[3] Hoogervorst, J.A.P., Dietz, J.L.G.: Enterprise Architecture in Enterprise Engineering, in: Enterprise Modeling and Information Systems Architecture, Vol. 3, No. 1, July 2008, pp 3-11, ISSN 1860-6059

[4] Hoogervorst, J.A.P., Enterprise Governance and Enterprise Architecture, Springer, 2009, ISBN 978-3-540-92670-2

[5] Barjis, J. (2007). Automatic Business Process Analysis and Simulation Based on DEMO. Journal of Enterprise Information Systems, Vol. 1, No. 4, pp. 365-381

based on a well-founded theory that integrates the notion of construction and operation of the enterprise in a fully implementation-independent way. Said approaches therefore result in unnecessary complex, unstable, and unwieldy models including not only the essential features of an enterprise. The same holds for the models of the supporting information systems, which are based on those enterprise models. In order to provide valuable information to business people who decide about requirements, use the solutions and decide about future strategies, both the enterprise models and the supporting information system models need to be provided on a high level of abstraction. Therefore, there is a need for new and innovative methodologies applying the notion of enterprise ontology, and for new methods transforming such ontological models into information system models[6]. The resulting information system models have a reference character. That means that they are stable since they are based on ontological models, which are completely implementation independent. A business domain is not going to change often, but the implementation of that business domain may change easily.

June 2009

Jan L.G. Dietz
Antonia Albani
Joseph Barjis

---

[6] Albani, A., Dietz, J., 2008. Software and Data Technologies, Second International Conference, ICSOFT/ENASE 2007, Barcelona, Spain, July 22-25, 2007, Revised Selected Papers. Vol. 22. Springer Verlag, Ch. Benefits of Enterprise Ontology for the Development of ICT-Based Value Networks, pp. 322.

Organization

The CIAO! and EOMAS workshops are organized annually as two international forums for researchers and practitioners in the general field of enterprise engineering. Organization of these two workshop and peer review of the contributions made to these workshops are accomplished by an outstanding international team of experts in the fields of enterprise engineering, modeling and simulation.

Workshop Chairs

CIAO! 2009

Jan L.G. Dietz   Delft University of Technology (The Netherlands)
Antonia Albani   Delft University of Technology (The Netherlands)

EOMAS 2009

Joseph Barjis   Delft University of Technology (The Netherlands)

Program Committee

CIAO! 2009

Wil van der Aalst   Aldo de Moor
Joseph Barjis   Hans Mulder
Bernhard Bauer   Moira Norrie
Emmanuel delaHostria   Martin Op 't Land
Johann Eder   Erik Proper
Joaquim Filipe   Gil Regev
Rony G. Flatscher   Pnina Soffer
Birgit Hofreiter   Pedro Sousa
Jan Hoogervorst   José Tribolet
Christian Huemer   Jan Verelst
Peter Loos   Robert Winter
Graham Mcleod

EOMAS 2009

Anteneh Ayanso   Ashish Gupta
Ygal Bendavid   Oleg Gusikhin
Tatiana Bouzdine-Chameeva   Johann Kinghorn
Manuel I. Capel-Tuñón   Fabrice Kordon
Samuel Fosso Wamba   Peggy Daniels Lee

Sponsoring Organizations

- SIGMAS (Special Interest Group on Modeling And Simulation
  of the Association for Information Systems)

- SIGSIM (Special Interest Group on Simulation
  of the Association for Computing Machinery)

Table of Contents

Modeling and Simulation

Enterprise Architecture and Governance

Enterprise Engineering – Applications

DEMO – Dissemination and Extension

Method Versus Model – Two Sides of the Same Coin?

Robert Winter, Anke Gericke, and Tobias Bucher

Institute of Information Management, University of St. Gallen,
Müller-Friedberg-Strasse 8, 9000 St. Gallen, Switzerland
{Robert.Winter,Anke.Gericke,Tobias.Bucher}@unisg.ch

Abstract. This article analyzes the state-of-the-art regarding the development of generic methods and reference models. The analysis shows that the related research disciplines, method engineering and reference modeling, tend to converge. Furthermore, it shows that the differentiation between generic methods and reference models should not be maintained because both artifact types feature activity-oriented elements as well as result-oriented elements. Depending on the artifact type, however, generic methods and reference models vary regarding the relative importance of the activity view and the result view. A generic problem solution (generic term for methods and reference models) can be interpreted as a sequence of activities which aim at the development of results. The insights into the commonalities among generic problem solutions provide the opportunity to define a unified design process in the field of design science research. Implications and unification challenges that are related to such a unified design process are presented at the end of the paper.

Keywords: Design Process, Method Engineering, Reference Modeling.

1 Introduction

Information systems (IS) researchers follow two main research approaches: behavioral research and design science research (DSR) [20, p. 76]. In contrast to behavioral research which is primarily aimed at advancing the body of knowledge through theory building, DSR is a problem solving paradigm which "has its roots in engineering" [20, p. 76]. The ultimate goal of the DSR approach is the development of useful artifacts that bear the potential to solve relevant IS problems [30, p. 253]. In this article, IS are generally understood as socio-technical systems. Socio-technical IS comprise all persons, business processes, software and information technology infrastructure that process data and information within an organization [cf. e. g. 2; 8; 38; 43].

March and Smith [30, p. 256 ff.] have established a widely accepted taxonomy of artifact types of DSR: constructs, models, methods and instantiations. In addition, design theories (relating to the design of artifacts, as opposed to general theories from the behavioral research paradigm) have been discussed as an extension of the DSR artifact taxonomy lately [cf. e. g. 25; 42].

Many European DSR communities are focusing on two specific artifact types: On the one hand, method engineering is addressing the development of generic methods and their adaptation in order to solve relevant IS problems. On the other hand, reference

A. Albani, J. Barjis, and J.L.G. Dietz (Eds.): CIAO!/EOMAS 2009, LNBIP 34, pp. 1–15, 2009.

modeling is aimed at the development of reusable conceptual models and their adaptation to solve relevant IS problems. However, as an analysis of contributions to the 2006 and 2007 International Conferences on Design Science Research in Information Systems and Technology (DESRIST) shows, research has mainly focused on the development of instantiations [10, p. 42]. Thus, we want to bridge this gap by analyzing the construction of generic methods and reusable conceptual models within their respective disciplines. Results of our analysis will represent the actual state-of-the-art regarding the development of both artifact types. Using this as a basis, topics for further research within both disciplines will be proposed. Addressing these research issues can contribute to the ongoing development of the method engineering and reference modeling discipline.

The remainder of the article at hand is structured as follows: In the following section, we present our analysis of related work on method engineering and reference modeling. The analysis shows that these two research disciplines are converging, in particular regarding "design knowledge". A convergence can also be observed in "artifact construction". These observations are outlined in the third section. They lead to the conclusion that generic methods and reusable conceptual models are two views on the same underlying object. This hypothesis is then taken up in the fourth section in which the relationship between generic methods and reusable conceptual models is analyzed in-depth. Within that section, a taxonomy of methods and models is presented. Based on that foundation, the fundamentals for a unified construction process for generic methods and reusable conceptual models are proposed. Since our hypothesis still holds true, we discuss some consequences for such a unified design process in the fifth section. The article closes with a summary and an outlook.

2   State-of-the-Art Analysis

Method engineering is concerned with the development of generic methods; reference modeling addresses the construction of reusable conceptual models. A review of the state-of-the-art in both disciplines is presented in the following. For this review the focus will be laid on the problem definition, construction/development and evaluation phases of the DSR process [cf. 32, p. 91 ff.].

2.1   Method Engineering

The method engineering (ME) discipline is concerned with the processes of constructing, adapting and implementing methods for the design of information systems [7, p. 276]. According to Brinkkemper, a method is "[…] an approach to perform a systems development project, based on a specific way of thinking, consisting of directions and rules, structured in a systematic way in development activities with corresponding development products" [7, p. 275 f.]. Such methods can be denoted as generic methods, as they are not restricted to solve only one specific problem, but rather address a class of (similar) design problems. In addition to this understanding of the term "generic method", different method definitions exist that particularly differ in respect to the method meta model [cf. 6]. All authors agree that a generic method consists of several activities and corresponding results [6, p. 1297]. Although activities and results are closely related to each other, they are often represented by two different models: activities and their sequence are represented by a procedure model while a result/deliverable model is used to represent results. Recently, so called process-deliverable diagrams [40, p. 36] have been proposed to jointly represent activities/activity sequences as well as results/deliverables and their relationships. A process-deliverable diagram is a combination of an UML activity diagram and an UML class diagram. In addition to activities and results (and the respective relationships), roles or techniques are often regarded as method meta model elements, too [6, p. 1297].

In order to be applicable for IS development, generic methods need to be adapted to the specific characteristics of the problem situation. This issue has been addressed in the ME discipline by proposing different construction processes for the development of so called situational methods [cf. e.g. 7; 21; 34; 39]. In order to provide a conceptual structure for these approaches, Bucher et al. [9, p. 35] and Bucher and Winter [10, p. 47 f.] suggest to differentiate situational method configuration and situational method composition. The distinguishing mark of situational method configuration is the adaptation of a so called base method against the background of a specific problem situation [9, p. 35]. By contrast, the fundamental idea of situational method composition is the selection and orchestration of method fragments with respect to the specifics of a problem situation [9, p. 35 f.]. Unlike situational method configuration, the composition process is not aimed at configuring one single base method, but at combining and aggregating several method fragments in order to establish new constructional results. Situational method composition is widely used and discussed in detail in the scientific literature [cf. e.g. 5, p. 6 f.].

Regarding these two different construction/development processes, the question arises how the problem situations can be characterized in which the methods will be used. Although the necessity for such a characterization of the problem situation (as part of the problem definition phase) has often been stated, there are only few approaches for defining a problem situation [9, p. 36]. Again, two different types can be differentiated. On the one hand, there are approaches that present different, predefined contingency factors such as "size of the project", "number of stakeholders" or "technology used" (cf. e.g. [24, p. 68 ff.] and [41], cited after [35, p. 12]). On the other hand, Bucher et al. [9] and Mirbel und Ralyté [31] characterize a situation e.g. by means of so called context type factors and project type factors [9, p. 37 ff.]. In contrast to the first approach, these factors are not predefined, but instead have to be identified individually for each and every problem situation and/or application domain.

Both the development of generic methods and the mandatory description of the problem situation have already experienced a wider research interest. In contrast, only few researchers have addressed the evaluation of methods up to now. Being the only contribution to this field to our knowledge, Pfeiffer und Niehaves [33, p. 5] present different evaluation approaches for the evaluation of methods such as case studies, action research or surveys.

2.2 Reference Modeling

Reference modeling is an IS research discipline dealing with the construction and application of reusable conceptual models, so called "reference models" [45, p. 48 ff.]. A reference model contains recommendations or references which can be used for the design of IS or the construction of other models [12, p. 35; 45, p. 48 f.; 46, p. 586 f.]. In addition to the reference character and to reusability (which are related to each other, cf. [44, p. 31 ff.]), there are other characteristics of reference models that are discussed in the literature as well. One of these characteristics is universality. Universality means that reference models should be valid solutions for an (abstract) class of problems [12, p. 35; 46, p. 584]. Over the past years, several procedure models have been developed that support the construction of reference models [cf. e.g. 15, p. 22 f.; 46, p. 591 ff.]. They do not differ significantly from each other and comprise the three generic construction phases outlined in section 2.

Up to now, only few contributions address the description of the problem situation in reference modeling [cf. e.g. 5, p. 7]. In contrast, numerous articles address the development phase of the construction process. For the reason of being adaptable to different problem situations when applying the reference model, the reference model has to be equipped with adaptation mechanisms during the development phase [45, p. 49]. Moreover, recommendations on how to adapt or how to use the reference model have to be provided [22]. Regarding the adaptation mechanisms, so called generating and non-generating approaches can be differentiated [17, p. 1]. Generating adaptation mechanisms are also referred to as configuration mechanisms and can be divided into (1) model type selection, (2) element type selection, (3) element selection, (4) synonym management and (5) presentation variation [3, p. 221 f.; 23, p. 136 ff.]. With respect to non-generating adaption mechanisms, aggregation, specialization, instantiation and analogy can be differentiated [cf. e.g. 44, p. 284 ff.; 45, p. 58 ff.]. After developing a reference model, an evaluation should be conducted in order to prove the utility of the model [13, p. 81]. In principle, such an evaluation can refer to the construction process itself or to the product of this process (i.e. the reference model). For both types of evaluation, different evaluation methods are available, such as the guidelines of modeling [37], ontological evaluation [14, 19] or evaluation based on case studies [13, p. 83]. The evaluation framework proposed by Fettke and Loos [13] provides an overview and systematization of different evaluation methods.

3   Convergence of Method Engineering and Reference Modeling

Following Hevner et al. [20, p. 87], different types of contributions can be differentiated in DSR: On the one hand, there are contributions in the area of "design knowledge" ("design construction knowledge" and "design evaluation knowledge"). On the other hand, "design artifacts" are considered as valid DSR contributions, too.

While analyzing the state-of-the-art of both disciplines, a convergence of both disciplines in respect of design knowledge can be observed. This is especially true for the area of design construction knowledge. Using this as a basis, we analyze whether such a convergence can be observed in the area of design artifacts as well. Thereafter, the conclusions drawn from these findings are presented, resulting in a proposed hypothesis that will be scrutinized in section 4.

3.1   Convergence in Respect of Design Knowledge

Researchers from the ME discipline [cf. 36; 47] as well as from the reference modeling discipline [cf. 4; 5] ask for the transfer of developed concepts to the respective "counterpart". Based there-on, several efforts have been undertaken to transfer existing research results in the different phases of the construction process, i.e. problem definition, development and evaluation. These efforts are presented in the following.

In reference modeling, only few approaches exist that deal with the specification of the problem situations in which a reference model should be used [5, p. 7]. By contrast, this topic has been addressed more intensively in ME. Although Schelp and Winter [36; 47] ask for the transfer of these results to the reference modeling discipline, contributions are still missing that describe how the specification of problem situations can be transferred to reference modeling in detail.

With respect to the transfer of adaptation mechanisms from reference modeling to ME (development phase), some first research results were achieved. Based on the assumption that both generic methods and reference models can be represented with the help of modeling languages, generating adaptation mechanisms (i.e. the configuration approach) have been formally transferred to generic methods [5, pp. 1, 7]. The applicability of element selection (one specific type of configuration) in ME could be shown on a procedure model as well [36, p. 569]. In this context, researchers still have to examine whether the other types of configuration (e.g. model type selection or element type selection) can also be applied to generic methods. On the contrary, the non-generating adaptation mechanisms instantiation and analogy have been considered scarcely in the ME discipline [4, p. 88; 5, p. 10].

The literature analysis (see section 2) shows that, although an evaluation is asked for in both research disciplines, this issue has hardly been addressed in research yet. That is why no contributions can be identified which deal with the transfer of evaluation approaches from one discipline to the other.

The analysis in respect of the convergence of both research disciplines in the area of design knowledge shows that the transfer of different approaches from ME to reference modeling and vice versa has already been done successfully or is at least intended. Using this as a basis, we analyze in the following whether or not this convergence can be observed for design artifacts, too.

3.2 Convergence in Respect of Design Artifacts

In order to determine whether the proposed convergence of ME and reference modeling can also be recognized regarding constructed artifacts, we will analyze case examples from current publications. For the identification of such case examples, we focus on an article of Bucher and Winter [10] that classifies contributions to the 2006 and 2007 International Conferences on Design Science Research in Information Systems and Technology (DESRIST) with respect to the type of artifact developed/presented. Based on this article, we select all articles that are classified as either method or model (see Table 1). However, we do disregard articles that are assigned to more than one type of artifact. We choose the article of Bucher and Winter [10] as well as the underlying DESRIST conference proceedings because this research community possesses a high research culture homogeneity – as they follow the DSR paradigm. Besides that, these proceedings enable us to take recent publications into account. Table 1 gives an overview about the chosen case examples.

Table 1. Case Examples

No.	Case Example	Predominant Type of Artifact
1	Arazy et al. 2006: Social Recommendations Systems: Leveraging the Power of Social Networks in Generating Recommendations [1]	Model
2	Gorla and Umanath 2006: On the Design of Optimal Compensation Structures for Outsourcing Software Development and Maintenance: An Agency Theory Perspective [16]	Model
3	Kunene and Weistroffer 2006: Design of a Method to Integrate Knowledge Discovery Techniques with Prior Domain Knowledge for Better Decision Support [26]	Method
4	Zhao 2006: Selective Encryption for MPEG-4 FGS Videos [48]	Method

Analyzing these case examples, it can be recognized that the development of (reference) models is predominant in the first two articles [see 1; 16]. In addition, both articles contain simple activity recommendations in the form of *Use the proposed model for the development of social filtering systems* [cf. 1, p. 320] or *Use the proposed model to optimize your compensation structures for outsourcing software development and maintenance* [cf. 16, p. 660]. These activity recommendations take the form of recommendations that are normally presented by generic methods. In contrast to articles one and two, generic methods are developed in articles three and four [see 26; 48]. Although activity recommendations are predominant, results (as normally presented with the help of reference models) are explicated as well, e.g. by giving examples for the results of some of the actions that are part of the method [cf. 26, p. 348 ff.; 48, p. 607 f.]. Summarizing this analysis, it can be stated that although one artifact type is always pre-dominant, aspects from both generic methods and reference models can be identified in all case examples simultaneously.

3.3 Intermediate Findings

In ME as well as in reference modeling, several topics such as artifact construction processes, contingency approaches for the adaptation of generic/reusable artifacts, mechanisms for adaptation, etc., have been developed separately (see section 2). Recently, both disciplines have increasingly cross-fertilized each other, resulting in the transfer of different concepts/topics from one discipline to the other (see above). This is not only true within the area of design knowledge of both disciplines. Rather, the convergence of ME and reference modeling can also be observed when looking at the actual construction of artifacts. Thus, our findings suggest that generic methods and reference models are somehow similar and/or related to each other. We therefore propose the following hypothesis:

> *Generic methods and reference models represent different views on the same underlying object.*

In the following, we justify this hypothesis using argumentative analysis. Our ultimate goal is to understand the relationship between generic methods and reference models. Insights into this relationship can serve as a foundation for a unified design process for both generic methods and reference models.

4 Discussion of the Hypothesis

In order to justify our hypothesis, we first introduce a taxonomy that compares generic methods and reference models. Based on that foundation, the relationship between generic methods and reference models is then explicated, allowing for the proposition of a unified design process.

4.1 Positioning Generic Methods and Reference Models in a Model Taxonomy

Based on the argumentation of Becker et al. [5, p. 1] that both generic methods (especially procedure models as constituent elements of methods) and reference models can be represented by means of models[1], we develop a taxonomy in which generic methods and reference models can be positioned.

IS models can be divided into description models, explanatory models and design models [27, p. 20]. Description models and explanatory models are understood as descriptive models whereas design models are considered to be prescriptive or instructional [28, p. 284]. The latter thus possess a recommendatory character. Due to the fact that generic methods as well as reference models express recommendations, these two model types are assigned to the category of design/prescriptive models. We will abstain from discussing descriptive models in the following.

In a second step, the prescriptive models can be further subdivided. A differentiation can be made regarding the way of recommendation: On the one hand, recommendations can refer to a (design) activity; on the other hand, they can refer to the result of that activity [28, p. 284]. Following this argumentation, generic methods/procedure models are assigned to the category of models that prescribe recommendations for activities. This is due to the fact that they provide instructions and recommendations about how to obtain a solution of an IS problem. On the contrary, reference models can be assigned to the category of models that represent a recommendation for a design result.[2]

In addition to generic methods and reference models, there are prescriptive models that are specific. This category of models includes e.g. project plans as specific activity recommendations or data and functional models of IT solutions as specific result recommendations. Those models have been exclusively developed for a single, specific

---

[1] In this context, the term "model" does not refer to a reusable conceptual model or the model term as defined by March and Smith [30]. Instead, it refers to the general meaning of a model as a representation or abstracted view of the underlying object.

[2] Reference models always express recommendations for design results. This is true for all reference models, irrespective of the reference model being e.g. a reference process model (action-oriented) or a reference model of an organizational structure (state-oriented). However, we will not focus on this content-related differentiation. Regardless of the reference model's design object being action-oriented (e.g. a process) or state-oriented (e.g. an organizational structure) it always provides a result recommendation for that design object.

problem. The distinguishing mark between generic methods and reference models on the one hand and those specific models on the other hand is the "intention for re-use". Generic methods and reference models possess a recommendatory character and are valid for an (abstract) class of design problems. Moreover, they are explicitly designed to be re-used. On the contrary, specific models express recommendations for the solution of one specific design problem only. They are not intended for re-use.

Fig. 1 summarizes the arguments and depicts the proposed taxonomy for prescriptive models.

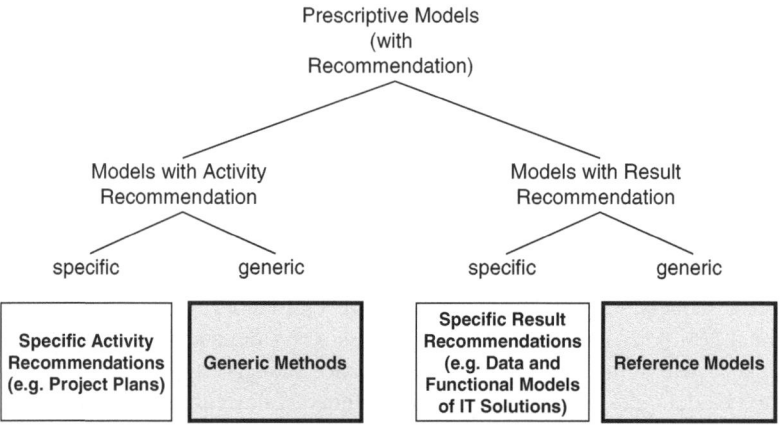

Fig. 1. Taxonomy of Prescriptive Models

4.2   Towards a Unified Design Process for IS

According to the taxonomy presented previously, generic methods and reference models can be differentiated primarily regarding their type of recommendation (activity vs. result). However, the literature analysis (see section 2) as well as the analysis in section 3 implicate that such a stringent differentiation cannot be maintained.

As outlined before, generic methods also describe possible results of the recommended activities [cf. e.g. 26; 48]. Similarly, reference models provide activity recommendations, e.g. on how to adapt the model to and/or on how to use the model in a certain problem situation [cf. e.g. 1; 16]. This argumentation leads to the conclusion that an activity view and a result view can be defined for both generic methods and reference models. Depending on the type of artifact (generic method vs. reference model), however, they vary regarding the relative importance that these two views have. Thus, we will denote each artifact that possesses both an activity and a result view as a problem solving artifact in the following.

Based on this argumentation, a problem solving artifact (or, rather, a problem solution) can be interpreted as a sequence of (partial) activities which develop (partial) results in order to solve a certain class of problems. Hence, problem solutions represent "means-ends relations" [11].

Fig. 2 illustrates this understanding of the problem solution process.

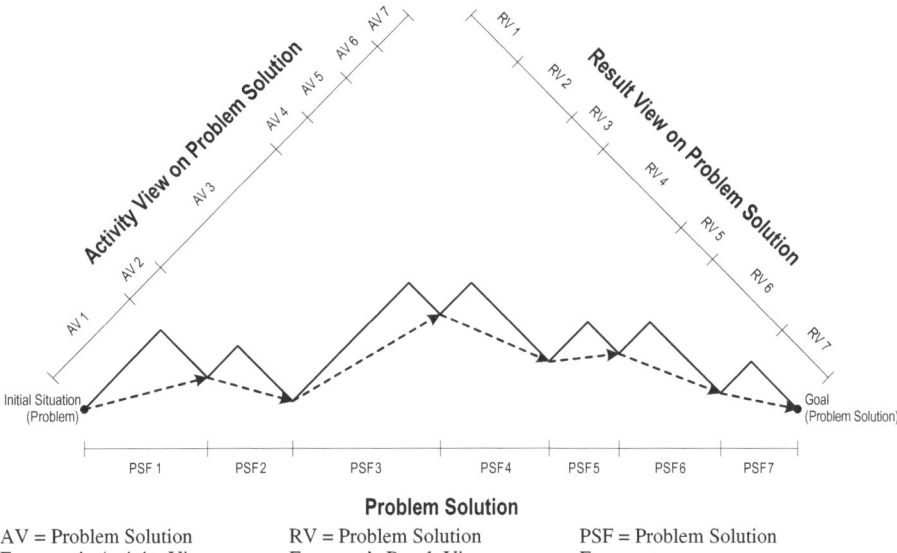

AV = Problem Solution
Fragment's Activity View

RV = Problem Solution
Fragment's Result View

PSF = Problem Solution
Fragment

Fig. 2. Problem Solution as a Sequence of Fragments that Have Both Activity and Result Character

Using this as a basis, we can differentiate two views on a problem solution:

- One problem solution view focuses on activities and can be designated as generic method. In detail, a generic method is understood as an artifact that represents a means-ends relation. The focus on activities inherent to this problem solution view is understood in the way that a generic method exactly describes how to create a solution for a problem/problem class, whereas the corresponding results are only implied or described rudimentarily.
- Another problem solution view focuses on results and can be designated as reference model. Equivalent to a generic method, a reference model is an artifact representing a means-ends relation. The focus on results inherent to this problem solution view is understood in the way that a reference model exactly characterizes a solution to a problem/problem class whereas the activities needed to solve the problem are only implied or described rudimentarily.

5   Consequences for a Unified Design Process

The arguments brought forward in the previous section support our hypothesis that generic methods and reference models are two sides of one and the same coin. This insight could and should lay the foundation for the definition of a unified design process for generic methods and reference models in DSR.

A similar development towards unified design can be observed in traditional engineering. In the 1960ies and 1970ies, different construction methods and processes have been developed in different research fields of the engineering discipline. Such "specific" construction methods and processes have been amalgamated into a "universal design theory" [cf. 18; 29] which is comprised of "findings and knowledge about design from different scientific and engineering disciplines in a consistent, coherent and compact form" [29, p. 203]. Domains integrated into the universal design theory are, for example, chemistry, chemical engineering, material science or technical biology.

Similar to the benefits outlined for the engineering discipline [29, p. 209], a unified design process for the construction of problem solutions in DSR would allow to achieve different benefits:

- The design process for the development of generic methods and reference models will become more efficient and reliable. This benefit can be realized by a design process that allows for the definition of construction processes, contingency approaches and mechanisms for artifact adaptation that are valid for both generic methods and reference models.
- If there is more than one possible tuple (activity, result) for a specific step in the problem solution process, the unified design process supports the evaluation and appreciation which of the possible tuples is more appropriate to contribute to the problem solution.
- Based on such a design process, different assumptions, procedures and outcomes of an artifact construction are made comparable and the design process becomes easier to control.
- With the definition of a unified design process, research efforts are reduced because several questions do not have to be discussed individually for each type of artifact.
- Based on such a unified design process, learning effects will be achieved that result from interdisciplinary knowledge acquisition. Furthermore, the construction of IS artifacts within interdisciplinary research teams is supported.

After explaining the advantages of a unified design process for generic methods and reference models, we will analyze the consequences with respect to the three construction phases: problem definition, development and evaluation (see section 2).

A problem solution, i.e. a generic method or a reference model, is used to solve relevant IS problems. As a precondition, it is necessary to describe the problem situation. In this context, two different assumptions about the consequences within a unified design process can be made: On the one hand, it can be reasonably assumed that existing problems are independent of the artifact type with which the problems are to be solved. Following this argumentation, research questions that address the description and specification of problems can be answered on a superordinate level for problem solutions. Hence, such research questions do not have to be considered individually in the ME discipline and in the reference modeling discipline. Instead, existing research results from both disciplines can be used to develop either a generic method or a reference model. This is true for research questions addressing the definition of problem situations in general (design knowledge), and problem descriptions for concrete application domains (design artifact). On the other hand, it can be assumed that different problem descriptions are necessary with respect to the kind of recommendations (activity/result) that will be developed. For the development of recommendations for activities, it might be necessary to describe the initial and the target state of a situation, whereas for the development of recommendations for results it is probably sufficient to describe only one state (either initial or target).

In the development phase of the construction process, research results in the area of design construction knowledge of generic methods and reference models can be used in both disciplines. Thus, research results regarding a procedure (sequence of activities) are not only valid for activity-oriented recommendations in the context of generic methods, but also for activity-oriented recommendations in the context of reference models. In turn, this is also true for result-oriented recommendations which are also valid in the context of both generic methods and reference models. Moreover, with the help of such a dedicated examination of the recommendatory character (activity vs. result), it will be considerably easier to use research results which have been gained in reference modeling with particular respect to results for the development of activities in ME instead. To make an example, the application of the adaptation mechanism "element type selection" of reference modeling to activities in ME could be introduced. Although such an application has already been conducted formally, the question arises under which terms and conditions this is possible for the content as well. For example, the utility of a method might be questionable if the element type "activity" was extracted from the method's procedure model. The clarification of such questions will be the basis for the construction of a unified design process. In addition to the consequences within the area of design knowledge, existing result-oriented recommendations could be integrated more frequently into the construction of activity recommendations and vice versa ("design artifact").

Finally, the consequences for a unified design process with respect to the evaluation of generic methods and reference models have to be analyzed. Since the research field of evaluation is not well-developed for both artifact types serious limitations have to be taken into account when making unification efforts. With respect to the design evaluation knowledge, it might be possible to use evaluation methods from ME and/or reference modeling to evaluate either a generic method or a reference model. Analogous to the development phase, we assume that evaluation results for activity recommendations are not only valid for generic methods, but also for reference models. This is also true for the evaluation of result recommendations for reference models that are also valid for generic methods ("design artifact").

6 Conclusion and Outlook

In the article at hand, we analyze the state-of-the-art of ME and reference modeling. These two disciplines of DSR for IS deal with the construction of generic methods and reference models, respectively. By analyzing the body of literature of both disciplines, a convergence of ME and reference modeling becomes evident. This is not only true in respect of the design knowledge of both disciplines, but also for the construction of concrete artifacts – as four case examples show. Thus, we propose the hypothesis that generic methods and reference models are two sides of one and the same coin. This hypothesis holds true as the argument can be brought forward that both generic methods and reference models can be viewed as a complex activity (procedure model, including all activities) as well as a complex result (including all intermediate/partial results). However, both artifact types vary with respect to their focus on the activity view and the result view, respectively. Following this argumentation, consequences for a unified design process have been presented. These arguments form the basis for further research activities.

In future research projects that deal with the development of generic methods and/or reference models, experience should be collected regarding the application of the design process of both ME and reference modeling. Those experiences will form the basis for the development of a unified design process that incorporates distinct parts that are valid for both artifact types as well as other parts that can only be applied under certain conditions.

Before developing such a unified design process, an in-depth analysis of the arguments brought forward in this article has to be performed, including formal analyses. In addition, the "artifact" term should be revisited as a consequence of our analysis. The strict definition of an artifact typology, as e.g. presented by March and Smith [30], might not be appropriate any more. Instead, a more general understanding of the term "artifact" should be developed. This can be achieved, for example, by defining a generic artifact as follows:

> *A generic artifact consists of language aspects (construct), aspects referring to result recommendations (model), and aspects referring to activity recommendations (method) as well as instantiations thereof (instantiation).*

In this respect, contributions should be called in that explicitly analyze the relations between these four aspects and put them in context to theories.

References

1. Arazy, O., Kumar, N., Shapira, B.: Social Recommendations Systems: Leveraging the Power of Social Networks in Generating Recommendations. In: Chatterjee, S., Hevner, A. (eds.) Proceedings of the 1st International Conference on Design Science Research in Information Systems and Technology (DESRIST 2006), Claremont, pp. 310–328 (2006)
2. Bacon, C.J., Fitzgerald, B.: A Systemic Framework for the Field of Information Systems. ACM SIGMIS Database 32, 46–67 (2001)
3. Becker, J., et al.: Configurable Reference Process Models for Public Administrations. In: Anttiroiko, A.-V., Malkia, M. (eds.) Encyclopedia of Digital Government, pp. 220–223. Idea Group, Hershey (2006)
4. Becker, J., Janiesch, C., Pfeiffer, D.: Reuse Mechanisms in Situational Method Engineering. In: Ralyté, J., et al. (eds.) Situational Method Engineering – Fundamentals and Experiences, pp. 79–93. Springer, Boston (2007)
5. Becker, J., et al.: Configurative Method Engineering – On the Applicability of Reference Modeling Mechanisms in Method Engineering. In: Proceedings of the 13th Americas Conference on Information Systems (AMCIS 2007), Keystone, pp. 1–12 (2007)
6. Braun, C., et al.: Method Construction – A Core Approach to Organizational Engineering. In: Haddad, H., et al. (eds.) Proceedings of the 20th Annual ACM Symposium on Applied Computing (SAC 2005), Santa Fe, New Mexico, USA, pp. 1295–1299 (2005)

7. Brinkkemper, S.: Method engineering: engineering of information systems development methods and tools. Information and Software Technology 38, 275–280 (1996)
8. Brookes, C.H.P., et al.: Information Systems Design. Prentice-Hall, Sydney (1982)
9. Bucher, T., et al.: Situational Method Engineering – On the Differentiation of "Context" and "Project Type". In: Ralyté, J., et al. (eds.) Situational Method Engineering – Fundamentals and Experiences, pp. 33–48. Springer, Boston (2007)
10. Bucher, T., Winter, R.: Dissemination and Importance of the "Method" Artifact in the Context of Design Research for Information Systems. In: Vaishnavi, V., Baskerville, R. (eds.) Proceedings of the Third International Conference on Design Science Research in Information Systems and Technology (DESRIST 2008), Atlanta, pp. 39–59 (2008)
11. Chmielewicz, K.: Forschungskonzeptionen der Wirtschaftswissenschaft. Schäffer-Poeschel, Stuttgart (1994)
12. Fettke, P., Loos, P.: Classification of reference models: a methodology and its application. Information Systems and e-Business Management 1, 35–53 (2003)
13. Fettke, P., Loos, P.: Multiperspective Evaluation of Reference Models – Towards a Framework. In: Jeusfeld, M.A., Pastor, Ó. (eds.) Proceedings of the Conceptual Modeling for Novel Application Domains Workshop (ER2003 Workshops ECOMO, IWCMQ, AOIS, and XSDM), Chicago, pp. 80–91 (2003)
14. Fettke, P., Loos, P.: Ontological Evaluation of Reference Models Using the Bunge-Wand-Weber Model. In: DeGross, J.I. (ed.) Proceedings of the Ninth Americas Conference on Information Systems, Tampa, pp. 2944–2955 (2003)
15. Fettke, P., Loos, P.: Der Beitrag der Referenzmodellierung zum Business Engineering. HMD – Praxis der Wirtschaftsinformatik 42, 18–26 (2005)
16. Gorla, N., Umanath, N.S.: On the Design of Optimal Compensation Structures for Outsourcing Software Development and Maintenance: An Agency Theory Perspective. In: Chatterjee, S., Hevner, A. (eds.) Proceedings of the 1st International Conference on Design Science Research in Information Systems and Technology (DESRIST 2006), Claremont, pp. 646–662 (2006)
17. Gottschalk, F., van der Aalst, W.M.P., Jansen-Vullers, M.H.: Configurable Process Models: A Foundational Approach. In: Lehner, F., et al. (eds.) Proceedings of the Multikonferenz Wirtschaftsinformatik 2006 (MKWI 2006), Passau (2006)
18. Grabowski, H., Lossack, R.-S., El-Mejbri, E.-F.: Towards a Universal Design Theory. In: Kals, H., van Houten, F. (eds.) Integration of Process Knowledge Into Design Support Systems: Proceedings of the 1999 CIRP International Design Seminar, University of Twente, Enschede, The Netherlands, Dordrecht, March 24-26, 1999, pp. 49–56 (1999)
19. Green, P., Rosemann, M.: Integrated Process Modeling: An Ontological Evaluation. Information Systems 25, 73–87 (2000)
20. Hevner, A.R., et al.: Design Science in Information Systems Research. MIS Quarterly 28, 75–105 (2004)
21. Karlsson, F., Ågerfalk, P.J.: Method configuration: adapting to situational characteristics while creating reusable assets. Information and Software Technology 46, 619–633 (2004)
22. Knackstedt, R.: Fachkonzeptionelle Referenzmodellierung einer Managementunterstützung mit quantitativen und qualitativen Daten – Methodische Konzepte zur Konstruktion und Anwendung, Doctoral Thesis, University of Münster, Münster (2004)
23. Knackstedt, R., Janiesch, C., Rieke, T.: Configuring Reference Models – An Integrated Approach for Transaction Processing and Decision Support. In: Manolopoulos, Y., et al. (eds.) Proceedings of the Eighth International Conference on Enterprise Information Systems (ICEIS 2006), Paphos, pp. 135–143 (2006)

24. Kornyshova, E., Deneckère, R., Salinesi, C.: Method Chunks Selection by Multicriteria Techniques: an Extension of the Assembly-based Approach. In: Ralyté, J., et al. (eds.) Situational Method Engineering – Fundamentals and Experiences, pp. 64–78. Springer, Boston (2007)

25. Kuechler, W.L., Vaishnavi, V.K.: Theory Development in Design Science Research: Anatomy of a Research Project. In: Vaishnavi, V.K., Baskerville, R. (eds.) Proceedings of the Third International Conference on Design Science Research in Information Systems and Technology (DESRIST 2008), Atlanta, pp. 1–15 (2008)

26. Kunene, K.N., Weistroffer, H.R.: Design of a Method to Integrate Knowledge Discovery Techniques with Prior Domain Knowledge for Better Decision Support. In: Chatterjee, S., Hevner, A. (eds.) Proceedings of the 1st International Conference on Design Science Research in Information Systems and Technology (DESRIST 2006), pp. 343–355. Claremont (2006)

27. Lehner, F.: Modelle und Modellierung in angewandter Informatik und Wirtschaftsinformatik oder wie ist die Wirklichkeit wirklich? Research Report No. 10, Chair of Business Information Science and Information Management, WHU – Otto Beisheim School of Management, Koblenz (1994)

28. Leppänen, M.: An Ontological Framework and a Methodical Skeleton for Method Engineering. Doctoral Thesis, University of Jyväskylä, Jyväskylä (2005)

29. Lossack, R.-S., Grabowski, H.: The Axiomatic Approach in the Universal Design Theory. In: Tate, D. (ed.) Proceedings of the First International Conference on Axiomatic Design (ICAD 2000), Cambridge, MA, pp. 203–210 (2000)

30. March, S.T., Smith, G.F.: Design and Natural Science Research on Information Technology. Decision Support Systems 15, 251–266 (1995)

31. Mirbel, I., Ralyté, J.: Situational method engineering: combining assembly-based and roadmap-driven approaches. Requirements Engineering 11, 58–78 (2006)

32. Peffers, K., et al.: The Design Science Research Process: A Model for Producing and Presenting Information Systems Research. In: Chatterjee, S., Hevner, A. (eds.) Proceedings of the First International Conference on Design Science Research in Information Systems and Technology (DESRIST 2006), pp. 83–106. Claremont (2006)

33. Pfeiffer, D., Niehaves, B.: Evaluation of Conceptual Models – A Structuralist Approach. In: Bartmann, D., et al. (eds.) Proceedings of the 13th European Conference on Information Systems (ECIS 2005), Regensburg (2005)

34. Ralyté, J., Rolland, C.: An Approach for Method Reengineering. In: Hideko, S.K., et al. (eds.) ER 2001. LNCS, vol. 2224, pp. 471–484. Springer, Heidelberg (2001)

35. Rolland, C.: A Primer for Method Engineering. In: Proceedings of the Informatique des Organisations d'Information et de Décision (INFORSID). Toulouse (1997)

36. Schelp, J., Winter, R.: Method Engineering – Lessons Learned from Reference Modeling. In: Chatterjee, S., Hevner, A. (eds.) Proceedings of the 1st International Conference on Design Science Research in Information Systems and Technology (DESRIST 2006), pp. 555–575. Claremont (2006)

37. Schütte, R., Rotthowe, T.: The Guidelines of Modeling – An Approach to Enhance the Quality in Information Models. In: Ling, T.-W., Ram, S., Li Lee, M. (eds.) ER 1998. LNCS, vol. 1507, pp. 240–254. Springer, Heidelberg (1998)

38. Tatnall, A., Davey, B., McConville, D.: Information Systems – Design and Implementation. Data Publishing, Melbourne (1996)

39. Tolvanen, J.-P.: Incremental Method Engineering with Modeling Tools: Theoretical Principles and Empirical Evidence. Doctoral Thesis, University of Jyväskylä, Jyväskylä (1998)

40. van de Weerd, I., Brinkkemper, S.: Meta-Modeling for Situational Analysis and Design Methods. In: Syed, M.R., Syed, S.N. (eds.) Handbook of Research on Modern Systems Analysis and Design Technologies and Applications, pp. 35–54. Idea Group, Hershey (2008)
41. van Slooten, K., Hodes, B.: Characterizing IS development projects. In: Brinkkemper, S., et al. (eds.) IFIP TC8 Working Conference on Method Engineering, pp. 29–44. Chapman & Hall, London (1996)
42. Venable, J.R.: The Role of Theory and Theorising in Design Science Research. In: Chatterjee, S., Hevner, A. (eds.) Proceedings of the 1st International Conference on Design Science in Information Systems and Technology (DESRIST 2006), pp. 1–18. Claremont (2006)
43. Vessey, I., Ramesh, V., Glass, R.L.: Research in Information Systems – An Empirical Study of Diversity in the Discipline and Its Journals. Journal of Management Information Systems 19, 129–174 (2002)
44. vom Brocke, J.: Referenzmodellierung: Gestaltung und Verteilung von Konstruktionsprozessen. Doctoral Thesis, University of Münster, Münster (2003)
45. von Brocke, J.: Design Principles for Reference Modelling. Reusing Information Models by Means of Aggregation, Specialisation, Instantiation, and Analogy. In: Fettke, P., Loos, P. (eds.) Reference Modelling for Business Systems Analysis, pp. 47–75. Idea Group, Hershey (2007)
46. von Brocke, J., Buddendick, C.: Reusable Conceptual Models – Requirements Based on the Design Science Research Paradigm. In: Chatterjee, S., Hevner, A. (eds.) Proceedings of the First International Conference on Design Science Research in Information Systems and Technology (DESRIST 2006), pp. 576–604. Claremont (2006)
47. Winter, R., Schelp, J.: Reference Modeling and Method Construction – A Design Science Perspective. In: Liebrock, L.M. (ed.) Proceedings of the 21st Annual ACM Symposium on Applied Computing (SAC 2006), pp. 1561–1562 (2006)
48. Zhao, H.: Selective Encryption for MPEG-4 FGS Videos. In: Chatterjee, S., Hevner, A. (eds.) Proceedings of the 1st International Conference on Design Science Research in Information Systems and Technology (DESRIST 2006), pp. 605–609. Claremont (2006)

Capturing Complex Business Processes Interdependencies Using Modeling and Simulation in a Multi-actor Environment

Jessica W. Sun, Joseph Barjis, Alexander Verbraeck, Marijn Janssen, and Jacco Kort

Jaffalaan 5, 2628 BX Delft, The Netherlands
w.sun@student.tudelft.nl,
{j.barjis,a.verbraeck,m.f.w.h.a.janssen}@tudelft.nl,
jacco.kort@mail.ing.nl

Abstract. Current business processes tend to become increasingly complex as a result of extensive interdependencies with partner organizations and the increasing use of technology for decision making in multi-actor environments. This complexity often grows to the extent that none of the involved actors is able to have a total overview of the complete end-to-end processes. An example of such a complex process is the application process of new merchants to obtain the possibility to accept electronic payments. Although static modeling of such a process can reveal valuable information about the structure and organization of business processes and the relation with the involved actors, a simulation model can provide more insight into behavior of the business system. With this knowledge the possible bottlenecks and problems within this process can be found, and then used to improve the business system resulting in an improved customer satisfaction. This paper describes the set-up of this simulation model and its use for finding efficient policy measures for involved actors.

Keywords: Animation, visualization, business process complexity, complex business process, multi-actor system, actors interdependency, business processes modeling, business process simulation, discrete event simulation.

1 Introduction

Nowadays business processes are often taking place in complex technological environments and multi-partner settings, where the business processes are for a large part depending on the performance of the underlying technology and relationship between the partners (Mintzberg, 1981). Since this technology is not always at hand within the organization that needs it, the outsourcing of technological solutions is becoming a standard way of working. For an outsourcing solution to work, access to external databases and feedback loops are often needed, which makes it crucial that computers can always interconnect in real-time after the architecture is finalized (Kaufmann & Kumar, 2008). The technological architecture does not stop at the organizational boundary and it enables interactions with customers and other businesses. In this way a complex constellation consisting of many different actors is created. These actors have

A. Albani, J. Barjis, and J.L.G. Dietz (Eds.): CIAO!/EOMAS 2009, LNBIP 34, pp. 16–27, 2009.

to cooperate with each other in order to make the processes work as intended. The performance of the complete process is dependent on the weakest link (Janssen, 2007). Therefore we need to analyze this problem on a network level and take the activities of the various actors into account. Only in this way the full, interconnected system can be analyzed. This means that information sharing between the actors of the subsystems should be maximized during the analysis or design of such a business system, since this increases the understanding of events within the inter-organizational processes and it improves the efficiency of decision making (Baffo et al., 2008).

On the level of a single organization this information often is the factor providing strategic advantage for companies, so the willingness to share this information will be low. Also the responsibilities of the individual companies just entail parts of the system, so their perspectives will differ from a holistic picture of the system and of the perspective of other actors. Therefore while the individual companies have a strong incentive to optimize their own organizational performance, they might not have incentives to further improve the overall system and its performance. Sometimes the perspectives are conflicting and optimization within one individual company will lead to a worsened system performance or undesired impacts elsewhere down in the process chain. Furthermore, the one who is paying for the investments might not be the beneficiary.

These types of complex multi-actor systems are difficult to analyze and possibilities for policy measures to solve the problems within these systems are therefore difficult to find. On top of it, trying to find policy measures from trial and error is often not possible, especially since the implementation of policy measures in one company could bring along unpredicted and sometimes unwanted results in the whole system and influence some of the other involved actors. An example of such a system in which the involved actors are both technically and organizationally interdependent is the electronic payment sector in the Netherlands. In the past the electronic payments sector was dominated by one party who intermediated all transactions among banks and businesses. There were many complaints about the costs for the merchants, and the Netherlands Competition Authority decided that this was an undesirable situation and it decided to reduce the monopoly position. The competition authority decided to introduce competition by splitting up the system into independent subsystems. Each subsystem should be provided by a number of providers, in this way stimulating competition. Thereby the influence of the end customer has also been increased since the customers can decide to choose a certain service provider. As a result the companies within the chain will have incentives to operate in a more efficient way, providing the end customer with higher quality products which costs less (Koppenjan, 2008).

When an individual company in a complex network wants to optimize system performance, it is very difficult for this company to analyze which policy measures will have an effect and what the exact effects will be. If a company in the investigated electronic payment system wants to increase the satisfaction for the end customer by enlarging the efficiency of the application process for merchants, it is difficult to predict whether a policy measure taken in his company will lead to less throughput time and less responsibilities and tasks for the merchant.

A popular method of analyzing complex and uncertain situation is using modeling and simulation (Carson, 2003). By analyzing the current situation with modeling and simulation tools, possible policy measures that will improve the system performance can be identified and quantitatively analyzed. In addition, by using the models in a strategic workshop, it will be possible to analyze the power structure within the business system and find the points where strategic behavior is possible, so the initiating company can take this behavior into account when making decisions about policy measures. This will be elaborated in the third section.

This paper reports a case study that involves a complex business setting. The main objective of this study is to investigate the possibilities to achieve system performance improvement in this multi-actor environment, by using the minimization of the throughput time of the business system as an assessment criterion. The study is carried out with a combined modeling and simulation method. This paper discusses mainly the static and dynamic modeling stage of the research and some first results of the modeling process.

2 Case: Electronic Payments Sector

In the case of the electronic payments sector, we have observed different objectives and different wishes (motivations) for making changes in the business system with different actors. Differences like this often lead to conflicting requirements for the business system and create a lot of confusion for the involved actors. This results in a situation in which multi-actor decision making is needed to solve the conflicts between the involved actors. The outcome of these types of decision making is often a situation in which all involved actors achieve parts of their goals but also have to give up some of their wishes. The final solution might focus just on the technical level, and lead to a situation in which the optimal business system performance will not be possible as a result of the many interfaces that are needed and the many points where mistakes can be made (Han et al., 2008). If this system is then put in operation, the result is often that the subsystems which are independently managed are not optimized for interacting with each other. This becomes especially visible for the end customer who might experience the mistakes when interactions among the subsystems fail. The problem is that this end customer does not have any knowledge on the cause of the problems. When the merchant discovers the causes of the problems, he is not in the position to handle the problem. When confronted with a problem from an end-customer in a complex network, companies often blame each other and do not provide a solution for the whole system as a solution goes beyond their organizational boundaries. Since within the chain there is no single actor that has the overview of the whole system, the involved actors will also not always be able to identify the cause of a problem. These types of situations are very difficult to handle, since the optimal solution is not easy to find (Sage, 2008). This could be due to for example that the considered causes of problems could be the wrong ones, or that the optimal system performance cannot be reached within the existing boundaries and requirements, even if the cause is known. This could lead to a situation in which the performance criteria are difficult to measure and the involved actors end up in a power struggle and arguing on responsibilities with each other. To improve the general system performance, the causes for problems and bottlenecks should be found, and arrangements should be made to prevent the power struggle and introduce performance measures to efficiently arrange the processes (Koppenjan & Groenewegen, 2005).

This case study is limited to the application process that is needed to acquire a working payment terminal to accept electronic payments. This system has technical complexity because of the interdependency between the technical systems, and organizational complexity since information sharing is needed for the system to perform. Also the competition authorities have introduced measures for competitiveness within the system to maximize the choice possibilities for the merchant. The process is analyzed from the viewpoint of the end customer, the merchant, and it contains the actions and events that have to be carried out before the terminal in the shop is operational.

In Figure 1, the application process is illustrated in a high level system diagram. As shown in Figure 1, the goal of the merchant is to have an efficient application process. This can be measured by a low throughput time of the application process, and few responsibilities for the merchant within the application process. This translates into the merchant requiring a short period of time between the time sending in a request for a new terminal and the moment when he has a working terminal within his shop, and a number of tasks for the merchant to be performed during the application process that is as low as possible. This means, less time is needed between an inquiry and the installation of the new terminal and simplified working process for merchant.

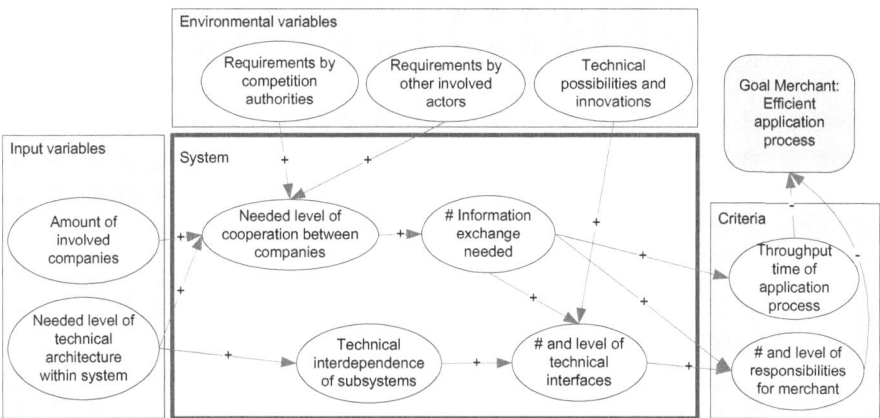

Fig. 1. System diagram for application process of merchants

The environmental variables in this system are the requirements set up by competition authorities in the form of regulation, requirements by other involved actors in the form of agreements, and the technical possibilities and innovations available at this moment. To fully maximize competition, the merchant should have fully independent choice possibility within the entire system. The other involved actors have set requirements to elements within the cards acquiring process and to let the system work. The technical possibilities and innovations currently enable the terminals to communicate with the acquiring processors over fast internet connections, both fixed and mobile. The term 'acquiring processors' refer to the non-bank service firms that handle electronic transactions of the customers, also called merchants, of an acquiring bank. Also it is technically possible for involved parties to update information into the acquiring processors' systems and databases in real-time making it possible to directly view the results of the updates. The input variables for this system are the number of involved parties within the application process and the level of the technical architecture within the system.

This case study provides an good example of modeling and simulation tools used in a complex multi-actor environment with technological interdependencies to provide more insight into the business processes and possibilities for improvement. As mentioned earlier, modeling and simulation enable actors to detect errors and potential problems within a business system in a cost effective manner (Ghosh, 2002). Especially for this case study for the identification of the relevant business processes, and the link to the operational performance of these business processes to strategic policy measures, a simulation model can be very helpful. (Greasly, 2000)

The crucial role of modeling within this research is to document the business processes as much as possible in a visualized way, to enable different parties to gain insight into the complexity and the potential solutions. For these reasons business process modeling of this system is conducted using rich graphical notations and diagrammatic languages. Creation or construction of business process models can help us to understand business processes, the actors involved, and to see the interdependencies between actors and complexity of processes (Shannon, 1998). For modeling to provide true value in this complex system, it is needed to look at the time-ordered dynamic behavior of the system. In this regard simulation plays a complementary role in understanding and analyzing complex systems (Zeigler et al, 2000). Simulation is a powerful tool for the analysis of new system designs, retrofits to existing systems and proposed changes to operational rules (Carson, 2003).

Currently there are five types of actors, including the merchant, who are crucial for making the application possible, and there is competition between actors for the major part of the system. On the technical level there is a need for communication between three of the five crucial parties to make the application possible. In Figure 2, the critical actors within the application process are shown. The merchant should provide information about his application choices to the terminal supplier, the acquirer and the telecom supplier. Then the application information is processed by the terminal supplier and the acquirer into the databases of the acquirer system. In addition to this, the application information should be inputted using the terminal. When the terminal and the databases of the acquiring processor contain the same information, the terminal can start accepting electronic payments. To achieve this, the terminal has to be able to communicate on a periodic basis with the acquiring system and the terminal management system through a telecom connection. The information exchange and the technical connections are also illustrated in Figure 2.

As shown in Figure 2, there are many moments when information exchange is needed between the involved actors. Since this information exchange occurs in a sequential order, it is important to know at what moment in the process which information exchange takes place. There are many possibilities for merchants to go through this process; therefore in this paper the situation for a very basic configuration of the application process will be further analyzed by a model of the process steps. Possible existing variations to this process (e.g. cooperation between certain

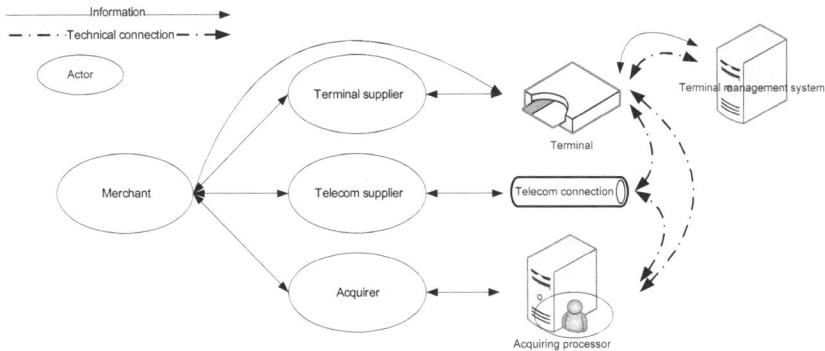

Fig. 2. Critical actors within the application process and their interaction

actors or added complexity as a result of coupled peripheral systems or internal networks) are not included in this. The results of the static and dynamic modelling of this basic application process are shown in sections 3 and 4.

3 Static Model of the Process Steps

The process starts with the merchant who wants to apply for a new terminal. As can be seen in Figure 2, the merchant should contact three parties, namely the acquirer, the terminal supplier and the telecom supplier. Basically the application processes can be carried out in parallel. The only restriction is that for the application of the acquirer the terminal IDs are needed. Therefore, the merchant first contacts the telecom supplier, and waits for the confirmation that the telecom supplier has activated a new telecom connection. Parallel to this the merchant contacts the terminal supplier. After the terminal supplier receives the application, the terminal supplier will then assign the IDs for the new terminals. Then the terminal supplier will send the terminal IDs and the terminal to the merchant. After the merchant has received the terminal, and has a confirmation that the telecom connection has successfully been set-up, the merchant can connect the terminal to the telecom connection. After the merchant received the terminal IDs, the terminal IDs can be entered into the application form of the acquirer and sent. The acquirer will handle the application forms they receive by inputting the information about the contract and about the terminal into the acquiring system. This information is needed for the acceptance of payments to be made on the terminal. After the acquirer has finished with this input, a letter is automatically generated and sent as a confirmation to the merchant. The parameters in this letter are needed by the merchant as input into the terminal. When the parameters in the terminal and the databases of the acquiring processor are matching, the terminal will be able to accept electronic payments. Also the terminal will update the terminal management system of the terminal supplier.

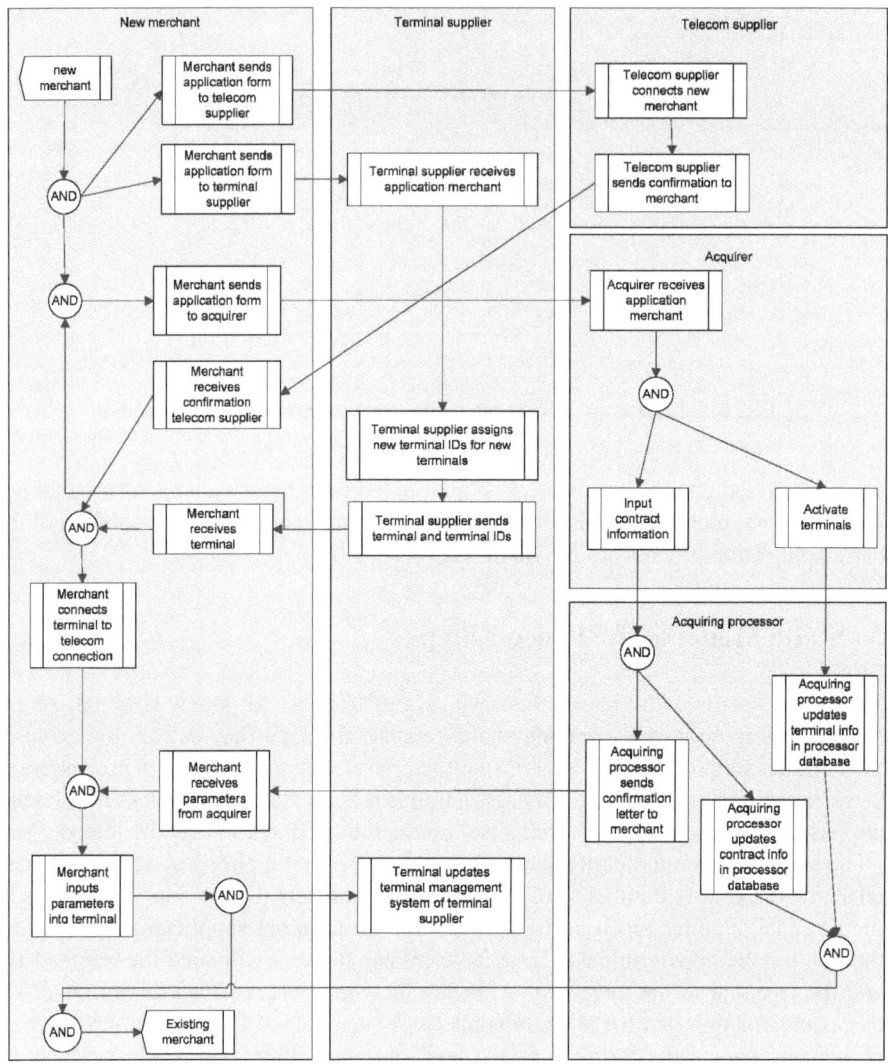

Fig. 3. Static model of application process for new merchants

It can be concluded that for a merchant to get a working terminal, there are multiple steps to complete and multiple tasks to perform. To perform the tasks the merchant is dependent on the performance of the other actors, especially for the provision of the information. From the static overview it becomes clear that there is a sequence in which the actions should take place, but it is not clear whether the moment the information needed by the merchant and the moment the information is provided to the merchant are corresponding. Also it is not clear whether the timing of information flows within the process between the other involved actors is optimal based on a static analysis.

4 Dynamic Model of the Process Steps

A dynamic model of the application process is made with the Arena software package (Kelton et al., 2002) to create an overview of the complex multi-actor system showing the time-ordered dynamics. In Figure 4, a screenshot of the model is shown, in which the application process of one merchant is analyzed with the estimated duration values for the times of the processes. This means for example for the package service, a value of one day is used, and for the update of information into the acquiring processor database a value of one hour is used. It should be mentioned that the values are not representing the actual values in the current situation, since between the different acquirers, terminal suppliers and telecom connection types there is such a variety of different values, it would involve too much data-analysis for this phase of the analysis. The model as it is only indicates the dynamic dependencies of the processes that are needed for the application processes, and do not yet contain the *exact* data for a quantitative analysis. This means that the outcomes provide a first indication based on estimates, which might not be applicable for each combination of actors. They do, however, provide a first insight into the dependencies and dynamics of the application process.

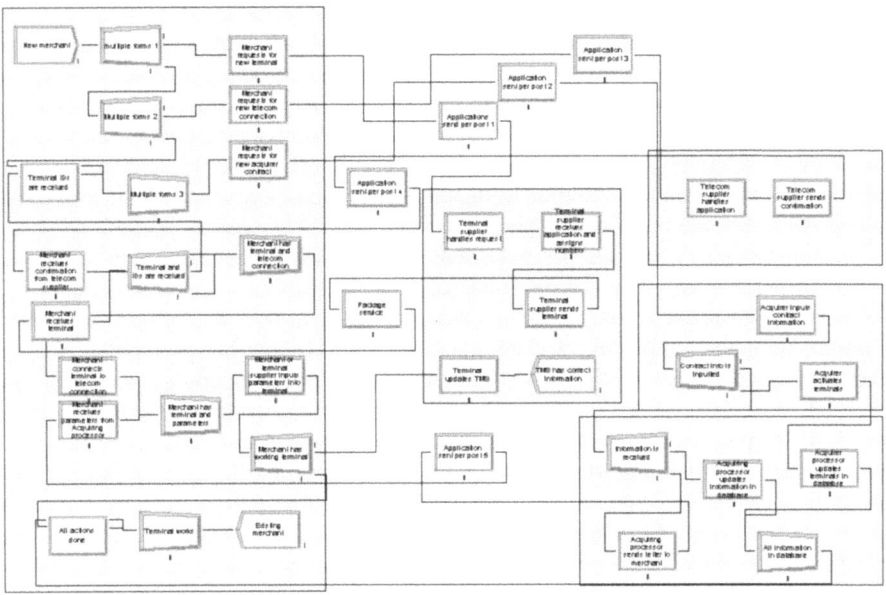

Fig. 4. Dynamic model of application process for new merchants (screenshot)

When running the simulation, it is observed that within the tasks for the merchant, there are four major points which may cause delays.

1. When the merchant waits for the terminal IDs, the merchant is not able to send in his application to the acquirer. In the current simulation it shows that this waiting time might take up to 145 hours.

2. When the merchant needs both the terminal and telecom connection, to connect the terminal to the telecom connection, a 24 hours delay may occur.
3. When the merchant has a terminal with telecom connection, and needs the parameters from the acquirer as input for the terminal, waiting time could be up to 25 hours.
4. The last point is when the data is already updated in the acquiring processor, but not yet manually inputted into the terminal. This again takes 25 hours.

Again, it should be mentioned that the values simulated (mentioned) are only to be considered as indications, resulting from the generic input values, and are not absolute outcomes. However, some first conclusions can be drawn from this dynamic model.

It can be assumed that the three actors delivering direct service to the merchant receive incentives from the merchant to optimize system performance. From the simulation model it can also be concluded that within this system it is very difficult, if not impossible, for an individual company to implement policy measures which will improve the system performance, since all actors involved need the cooperation of other actors to achieve a better situation for the end customer. This suggests that agreements between multiple parties could lead to a better situation for the end customer. The points within the system where cooperation is needed between actors are illustrated in Figure 5.

From the acquirer's perspective, it can also be assumed that the acquirer has interest in a low throughput time from the point the contract is received from the merchant to the moment the terminal can start accepting electronic payments. As can be learned from the simulation, it is then important for the acquirer that the acquiring system updates the information of the new merchant before the merchant has had the chance to input the parameters into the terminal. In addition, it can be seen that for the acquirer it is important that the merchant receives the terminal IDs as soon as possible after the application for new terminals has been sent to the terminal supplier, since that is the moment when the merchant can fill in the forms needed for the acquirer contract.

For the perspective of the terminal supplier, it is clear that it is important that the acquirer sends the parameters that the merchant should input in his terminal as soon as possible to the merchant. This is important for the terminal supplier who input the terminals on behalf of the merchants, so they can do this as soon as possible, most preferable at the same moment the terminal is delivered, which reduces their operational costs. Another important point for the terminal supplier is that the telecom connection is activated successfully before the terminal is delivered. This is equally important for some of the terminal suppliers that install the terminals for the merchants, since the telecom connection might be needed to input the parameters.

Based on the static and dynamic analysis, the involved actors within the application process can be divided into two groups. On the one hand we have the acquirer and the telecom supplier, and on the other hand the other parties, including the merchant. Within the current configuration of processes, the acquirer and the telecom supplier will have a strategic advantage, since the other actors have an interest in their cooperation. The acquirer and terminal supplier are the actors who rely the most on the other actors for a good system performance. What also can be concluded is that since the acquirer and terminal supplier are mutually dependent on each other for a good performance, this offers an opportunity to restructure the process in such a way

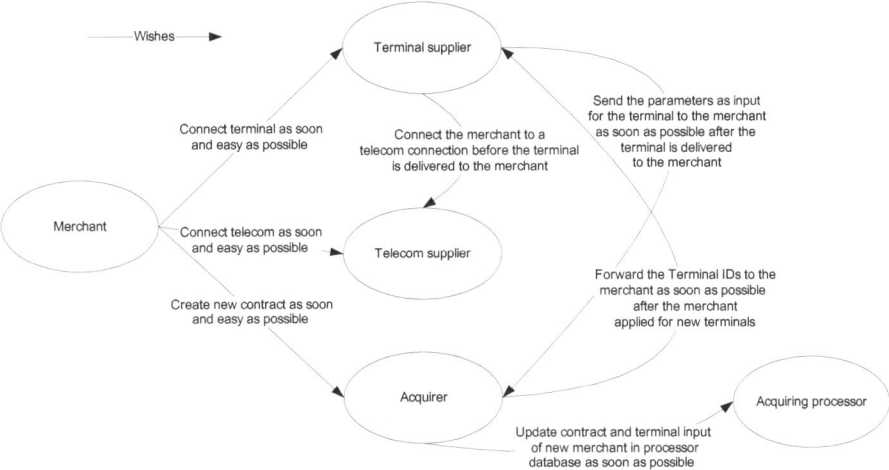

Fig. 5. Possible wishes (improvements) between involved actors

that both parties can find advantages. Since both actors will have an interest in this cooperation, this could be the most feasible way to improve the system performance.

5 Conclusion

In a more network focused economy, actors become more dependent on each other. These dependencies easily result in business failure, which needs to be analyzed beyond the individual organizational boundaries. In this paper, using an example from the electronic payments sector, it was demonstrated how modeling and simulation tools can be effectively used to find policy measures in multi-actor environments with technological interdependencies. The simulation models shows that there are many policy measures possible for improving business system performance, for example, a shorter throughput time to achieve better customers' satisfaction. Using the static and dynamic models, it can be concluded that when one company optimizes their own business processes, this might be suboptimal and will not immediately result in a better total system performance. There are two actors within the application process which are strongly dependent on each other, namely the acquirer and the terminal supplier. If one of these two companies wants to implement policy measures, it is crucial for them to make good arrangements with each other to obtain an improvement in the total system performance. Since they are so strongly interdependent, finding cooperation possibilities within the chain of actors that will lead to minimal strategic behavior from one of the actors has the most chance of success. In this system a restructuring of the application process by cooperation between the acquirer and terminal supplier could significantly reduce the throughput time. For the performance of the system, the acquirer and the terminal supplier still stay dependent on the other two actors in the process, namely the acquiring processor and the telecom supplier, to work along. Since the latter two actors are less dependent on other actors in this application process, the acquirer and terminal supplier should try and find incentives for these two actors to cooperate. This will probably be more difficult since in this case the sense of urgency is not divided evenly.

The main goal of this research was also to investigate the potential of using simulation and modeling as a method to improve performance of a complex network consisting of many, interdependent actors We have demonstrated that such a complex system indeed needs other policy measures than optimization within one company, and that the simulation model was able to show where to find the interdependencies between the companies and how this could affect the policy measures. More researches are needed to provide recommendations in using simulation and modeling to improve business system performance, however, the modeling process presented in this study may be applicable to a comparable process.

6   Recommendations

To further elaborate the research described in this paper, there are a number of fruitful research directions. For example further research about the actual times the processes take in the real life is a good option to start with, this way the current behavior can be taken into account. Discrete-event simulation tools like Arena are very suitable for this application since it involves a business system which involves many queues within the network, and the activities are distributed irregularly in time (DeBenedictis et al., 1991). Further research about the variance within the data can also be interesting, to analyze how lean the process is and which implications this has for the system performance. Also, further research could be done into the input of the actual numbers of merchant and capacities of the companies to see variations in different companies.

Approaching this business system from another perspective, it could be interesting to find out the wishes and perspective of the merchant, to figure out which values of throughput times and number of tasks are acceptable. Interviews and data analysis are needed for this research.

Finally, if a company has the wish to optimize its system performance, further research about the effects of policy measures of companies could be conducted, to find out how these could improve the overall system behavior. Instead of finding these possible improvements manually, it is also becoming increasingly common to couple simulation models to optimization tools that will calculate the optimal parameters within the business system. By combining a simulation model with the current values with such an optimization tool the decision making can be supported and improved for the entire chain of actors (Wiedemann & Kung, 2003). It must be aware though that in this case the opportunistic behavior of actors will obviously be very difficult to incorporate in such an optimization model.

References

1. Baffo, I., Confessore, G., Galiano, G., Rismondo, S.: A multi agent system model to evaluate the dynamics of a collaborative network. In: ICEIS 2008 – Proceedings of the 10th International Conference on Enterprise Information Systems, AIDDS, pp. 480–483 (2008)

2. Carson II, J.S.: Introduction to modeling and simulation. In: Proceedings of the Winter Simulation Conference, pp. 7–13 (2003)
3. DeBenedictis, E., Ghosh, S., Yu, M.-L.: A Novel Algorithm for Discrete-Event Simulation. IEEE Computer 24(6), 21–33 (1991)
4. Ghosh, S.: The Role of Modeling and Asynchronous Distributed Simulation in Analyzing Complex Systems of the Future. Information Systems Frontiers 4(2), 161–177 (2002)
5. Greasley, A.: Effective Uses of Business Process Simulation. In: Proceedings of the 2000 Winter Simulation Conference, pp. 2004–2009 (2000)
6. Han, K., Kauffman, R.J., Nault, B.R.: Relative importance, specific investment and ownership in interorganizational systems. Information Technology Management 9, 181–200 (2008)
7. Janssen, M.: Adaptability and Accountability of Information Architectures in Interorganizational Networks. In: International Conference on Electronic Governance (ICEGOV 2007), Macao, December 10-13, 2007, pp. 57–64 (2007)
8. Kauffman, R.J., Kumar, A.: Network effects and embedded options: decision-making under uncertainty for network technology investments. Information Technology Management 9, 149–168 (2008)
9. Kelton, W.D., Sadowski, R.P., Sadowski, D.A.: Simulation with Arena, 2nd edn. McGraw-Hill, New York (2002)
10. Koppenjan, J., Groenewegen, J.: Institutional design for complex technological systems. International Journal of Technology, Policy and Management 5(3), 240–257 (2005)
11. Koppenjan, J.: Creating a playing field for assessing the effectiveness of network collaboration by performance measures. Public Management Review 10(6), 699–714 (2008)
12. Mintzberg, H.: Organizational Design, Fashion or Fit? Harvard Business Review 59(1), 103–116 (1981)
13. Sage, A.P.: Risk in system of systems engineering and management. Journal of Industrial and Management Optimization 4(3), 477–487 (2008)
14. Shannon, R.E.: Introduction to the art and science of simulation. In: Proceedings of the 1998 Winter Simulation Conference (1998)
15. Wiedemann, T., Kung, W.: Actual and future options of simulation and optimization in manufacturing, organization and logistics. In: Verbraeck, A., Hlupic, V. (eds.) Proceedings 15th European Simulation Symposium, SCS Europe, pp. 627–637 (2003)
16. Zeigler, B.P., Praehofer, H., Kim, T.G.: Theory of modeling and simulation, 2nd edn. Academic Press, London (2000)

A Heuristic Method for Business Process Model Evaluation

Volker Gruhn and Ralf Laue

Chair of Applied Telematics / e-Business*
Computer Science Faculty, University of Leipzig, Germany
{gruhn,laue}@ebus.informatik.uni-leipzig.de

Abstract. In this paper, we present a heuristic approach for finding errors and possible improvements in business process models. First, we translate the information that is included in a model into a set of Prolog facts. We then search for patterns which are related to a violation of the soundness property, bad modeling style or otherwise give raise to the assumption that the model should be improved. By testing our approach on a large repository of real-world models, we found that the heuristic approach identifies violations of the soundness property almost as accurate as model-checkers that explore the state space of all possible executions of the model. Other than these tools, our approach never ran into state-space explosion problems. Furthermore, our pattern system can also detect patterns for bad modeling style which can help to improve the quality of the models.

1 Introduction

In the past years, numerous static code analysis tools have been developed for finding software bugs automatically. They analyze the source code statically, i.e. without actually executing the code. There exist several good tools that are matured to be useful in production environments. Examples of such tools are Splint (`www.splint.org`), JLint (`jlint.sourceforge.net`) or Find-Bugs (`findbugs.sourceforge.net`).

These static analysis tools use very different technologies for localizing errors in the code, including dataflow analysis, theorem proving and model checking [1]. All the tools and technologies have in common that they use some kind of **heuristics** in order to find possible problems in the code. This means that it can happen that such a tool reports a warning for code that is in fact correct (*false positive*) as well as the tool can fail to warn about an actual error (*false negative*).

It is important to mention that static analysis can be applied not only for locating bugs, but also for detecting so-called "bad code smells" like violations of coding conventions or code duplication. This means that not only erroneous code but also code that is hard to read or hard to maintain can be located.

---

* The Chair of Applied Telematics / e-Business is endowed by Deutsche Telekom AG.

A. Albani, J. Barjis, and J.L.G. Dietz (Eds.): CIAO!/EOMAS 2009, LNBIP 34, pp. 28–39, 2009.
© Springer-Verlag Berlin Heidelberg 2009

In this paper, we show how some ideas behind such static analysis tools can be transferred to the area of business process modeling.

2 The EPC Notation

We have developed our approach using the business process modeling language Event-Driven Process Chains [2]. There are two reasons for this choice: The first reason is that this modeling language is very widespread (at least this is the case for Germany) and we have been able to collect a large repository of models. Secondly, the EPC notation is a rather simple notation which is made up of the basic modeling elements that can be found in more expressive languages like BPMN or YAWL as well. These basic constructs will be introduced in the remainder of this section.

EPCs consist of functions (activities which need to be executed, depicted as rounded boxes), events (pre- and postconditions before / after a function is executed, depicted as hexagons) and connectors (which can split or join the flow of control between the elements). Arcs between these elements represent the control flow. The connectors are used to model parallel and alternative executions. There are two kinds of connectors: Splits have one incoming and at least two outgoing arcs, joins have at least two incoming arcs and one outgoing arc.

AND-connectors (depicted as ⊗) are used to model parallel execution. After an AND-split, the elements on all outgoing arcs have to be executed in parallel. An AND-join connector waits until all parallel control flows that have been started are finished.

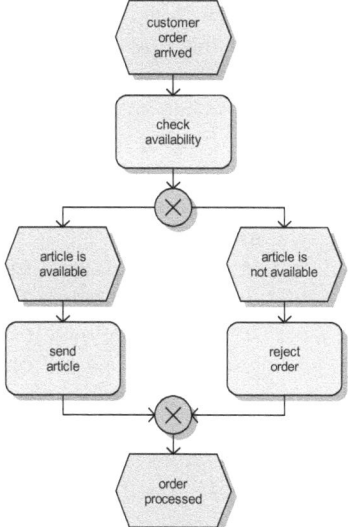

Fig. 1. Simple Business Process modeled as EPC

XOR-connectors (depicted as ⓧ) can be used to model alternative execution: An XOR-split has multiple outgoing arcs, but only one of them will be processed. An XOR-join waits for the completion of the control flow on the selected arc.

Finally, OR-connectors (depicted as ⓥ) are used to model parallel execution of one or more flows. An OR-split starts the processing of one or more of its outgoing arcs. An OR-join waits until all control flows that have been started (usually by a corresponding OR-split) are finished.

Fig. 1 shows a simple business process modeled as EPC diagram. The meaning of this model is as follows: When a request from a customer arrives, the availability of the product has to be checked. If it is available, the item will be sent; otherwise the customer will get a negative reply.

3   Our General Approach: Pattern Matching

The key idea of our approach is to search for patterns of "bad modeling". For the purpose of finding such patterns in a model, we used logic programming with Prolog. The XML serialization of the model has been translated into a set of facts in a Prolog program (as described in [3]). Each function, event, connector and arc in the model is translated into one Prolog fact.

Furthermore, we have constructed Prolog rules that help us to specify the patterns we are looking for.

We started with defining some rules describing the basic terminology, for example by specifying that a connector is called a split if it has one incoming and more than one outgoing arc or by recursively defining what we want to call a path from some node to another.

Secondly, we defined some important relations between split and join nodes that we can use to define the patterns. The most important definition is the one for the relation match(S,J). It means that a split S corresponds to a join J such that S branches the flow of control into several paths that are later merged by a join J. As we cannot assume that splits and joins are properly nested, this definition is the prerequisite for finding patterns that are related to control-flow errors in arbitrary structured models. We have defined the Prolog clause match(S,J) such that S is a split, J is a join and there are two paths from S to J whose only common elements are S and J.

Furthermore, we defined exits from and entries into a control block between a split S and a join J for which match(S,J) holds. For example, an exit from such a structure is defined such that there is a path from S to an end event (i.e. an event without outgoing arc) that does not pass J or a path from S to S that does not pass J. In Fig. 2, model (c) is the only one that has an exit from the control block between s and j.

By a systematic analysis of all possible patterns of matching split-join pairs with or without "exits" and "entries", we developed a set of 12 patterns that are indicators for control-flow errors.

Here, we will discuss one such pattern which is a good example of the heuristic nature of our approach: One of our rules states that we suppose the existence of

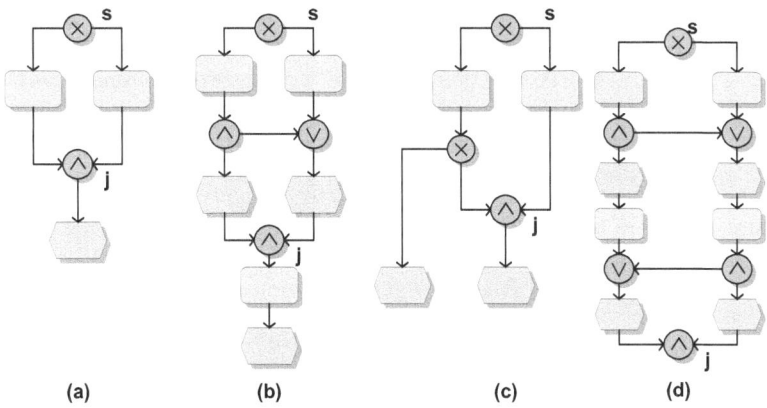

Fig. 2. The rightmost model fragment does not have a deadlock, the others have

a deadlock if there is an (X)OR-split s and an AND-join j such that `match(s,j)` holds, regardless of whether there are exits or entries between the split and the join. In most real-world models, such a pattern is indeed related to a deadlock, some cases are shown in Fig. 2 (a)-(c).

In these models, the outgoing arcs from the XOR-split s are later joined by an AND-join j. While only *one* outgoing flow from the XOR-split will be processed, the AND-join has to wait until *all* incoming arcs have been completed - a typical deadlock situation.

However, in some rare occasions (as the one shown in Fig. 2 (d)) the pattern is found in a model that in fact does not have a deadlock.

The existence of such rare cases where our rules would give the wrong information is inherent to the heuristic idea of our approach. It was not our goal to make our pattern catalogue as complete as possible. In the same way as it is known from code analysis tools, we had to find a balance between accuracy (i.e. prevention of false positives and false negatives) and speed of execution[1].

In the next sections, we will show which kind of problems can be located by applying different kinds of patterns.

4   Control-Flow Errors

The most important correctness criterion for business process models is the soundness property, originally introduced by van der Aalst for workflow nets [4,5] and later adapted to the EPC notation [2,6].

For a business process model to be sound, three properties are required:

1. In every state that is reachable from a start state, there must be the possibility to reach a final state *(option to complete)*.

---

[1] In fact, the case shown in Fig. 2 (d) is even considered by the latest version of our Prolog rule set: The rules will produce an "error" alert for the models (a)-(c) and a "possible error" alert for model (d).

2. If a state has no subsequent state (according to the transition relation that defines the precise semantics), then only events without outgoing arcs (end events) must be marked as being "active" in this state *(proper completion)*.
3. There is no element of the model that is never processed in any execution of the model *(no needless elements)*.

Violations of the soundness criterion usually indicate an error in the model. Therefore, 12 out of the 24 patterns we have defined so far aim to locate control-flow errors that can lead to a violation of the soundness property. An example (the combination of an (X)OR-split and an AND-join) has already been discussed in Sect. 3.

5   Comprehensibility and Style

Correctness (in terms of the soundness property) is not the only quality requirement for business process models: One of the main purposes of such models is to be used as a language in a discussion between humans. In particular, the models can serve as a bridge between the stakeholders in a software development project. They are formal enough to serve the demands of software developers but easy enough to be understood by business experts as well.

For this purpose, business process models should be as easy as possible to comprehend. If there is a choice among different modeling elements to express the same situation, the most comprehensible alternative should be used.

For example, in some cases it is possible to replace an OR-connector by an AND- or XOR-connector which describes the situation much better (for a human reader) without changing the semantics of the model.

As example, take a control block where an AND-split starts two paths that are executed in parallel. Formally, it is correct to join both paths using an OR-join. The meaning of the OR-join is to wait for all paths that have been started as a prerequisite for transferring control to its outgoing path. This means that in Fig. 3 (a), the OR-join acts exactly as an AND-join. While it would not make a difference for the actual meaning of the model, the readability of the model can be improved by substituting the OR-join by an AND-join. The same idea can be applied for the other model fragments in Fig. 3: In model (b) and (c), the

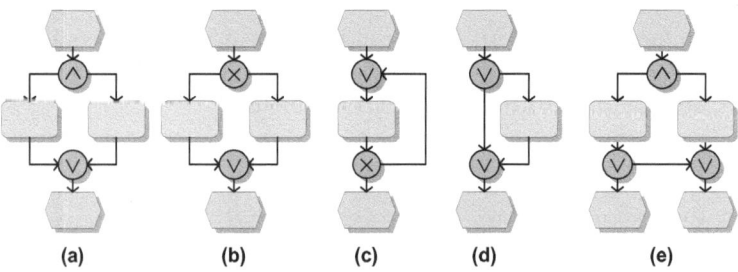

Fig. 3. Models with OR-connectors that should be replaced

OR-join should be replaced by an XOR-join. In model (d), a pair of XOR-split and XOR-join should be used to model the fact that an activity can either be skipped (by taking the left path) or executed (by taking the right one). Fig. 3 (e) shows another situation where an OR-split can be replaced by an XOR-split. With an XOR-split, the mental load for the reader of the model is reduced: He or she has not to consider the case that both outgoing arcs of the split are followed.

Using Prolog rules that specify the above patterns (and a few more that are not described here due to space restrictions), we are able to advice the modeler to change the model in order to improve its readability.

An organization can add own style rules, for example in order to enforce the policy that all pairs of splits and joins have to be properly nested (which is not required by the notation, but sometimes desirable).

6   Pragmatic Errors

So far, we have discussed "technical" errors (like deadlocks) and the readability of the model. It is almost impossible to validate automatically whether the model really represents the real business process without building an ontology of the application domain (see for example [7,8]) which is usually far too time-consuming and expensive.

There are however, some patterns that "look like" the model does not reflect the real business process. In such a situation, the modeler can be informed to double-check the questionable part of the model. We have discussed one such case (that we call Partial Redo pattern) in [9]. Another (simpler) pattern is shown in Fig. 4: After a function has been performed, an XOR-connector splits the flow of control and exactly one of two alternative events happens. However, afterwards both paths are joined again and the future execution of the process is the same regardless of which event actually occurred. But why does the model show that two different events can happen if the execution actually does not care whether the right event or the left event occurred? In some cases, this might make sense for documentation purposes, but it is also not unlikely that the modeler did forget something in the model.

In our survey of real-world models we found examples where this pattern indeed was the result of an error that had to be corrected, but of course we

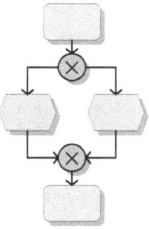

Fig. 4. Why is the alternative between two events modeled if these events do not have an effect on future execution?

also found models in which such a pattern just helps to understand the possible progress of the business process. The false warnings produced by the heuristics are less problematic, compared to the reward when a correct warning helps to locate an actual modeling error.

7 Validation

We searched for the patterns described above (and some more that cannot be described in detail due to space restrictions) in a repository of 984 models. Those models have been collected from 130 sources. These sources can be categorized as follows:

- 531 models from the SAP R/3 reference model, a widespread business reference model
- 112 models from 31 bachelor and diploma thesises
- 25 models from 7 PhD thesises
- 13 models from 2 technical manuals
- 82 models from 48 published scientific papers
- 12 models from 6 university lecture notes
- 4 models from sample solutions to university examination questions
- 88 models from 11 real-world projects
- 88 models from 7 textbooks
- 29 models from 14 other sources

Among the models in our repository, there is a great variation in size of the models, purpose of modeling, business domain and experience of the modelers. For this reason, we think that the models represent a reasonable good sample of real-world models.

Before doing the soundness analysis with these tools, we reduced all models using soundness-preserving reduction rules as described in [10,6]. 534 of the models have been reduced to a single node which means that they are sound. An analysis was necessary for the remaining 450 models.

In order to compare our heuristic results to the results of exact soundness analysis, we selected three well-known open-source tools that check business process models for the soundness property: *EPCTools*, the *ProM plugin for EPC soundness analysis* and the *YAWL Editor*. These tools have to compute the state space of all possible executions of the model. Because this state space can contain a great number of states, such tools can suffer from the so-called state-space explosion - a situation in which there is not enough available memory to store all possible states or the computation can not be done in a reasonable time.

All these tools run as a Java program. We executed the tools on an Intel Core2 Duo CPU running at a speed of 3 GHz. By starting the Java virtual machine with the option -Xmx1536m, we allowed a heap size of 1.5 GB to be used.

EPCTools [11,12] calculates a fixed-point semantics for a model. If such a fixed-point semantics exists, a temporal model-checker is used by *EPCTools* for deciding about the soundness property. For the majority of models, an analysis result was given in a few seconds. There was, however, one model for which the analysis took more than 10 minutes; this model was validated in 63 minutes. For 7 models, *EPCTools* had to stop the computation because of an *Out of Memory* error.

The *ProM plugin for EPC soundness analysis* [13] uses the semantics defined by Mendling [6] for constructing a transition system for the model. For 31 models, ProM failed to deliver a result because of an *Out of Memory* error. For 5 models, the computation took more than 10 minutes, the longest computation time was 26 minutes.

The third tool, *YAWL Editor* [14,15], originally has been constructed for analyzing YAWL models. While the mapping of EPC modeling elements to YAWL is straightforward, there is an important difference between EPC and YAWL: YAWL does not support process models with more than one start node. In order to avoid the problems that arise from the different instantiation semantics for EPC and YAWL models [16], we considered only those non-trivial 203 models for which the EPC model has exactly one start event (after applying the reduction rules). YAWL Editor has a built-in restriction that stops the execution of the analysis if the state-space exceeds 10,000 states. This is necessary, because the YAWL semantics allows an infinite state space [14]. This restriction was enforced for 13 models, meaning that no analysis result for them was available. While the computation was very fast for the majority of the models (133 have been analyzed in less than 1 second), there were also some that took much longer: For 8 models, the computation took more than 10 minutes. Two models could not be analyzed within one hour; the longest computation time was 390 minutes.

For all tools which we have tested, some models turned out to be "hard cases" where a Java heap space of 1.5 GB and a time of one hour was not enough to judge about the soundness property by exploring the state space. EPCTools had 8 such "hard cases" (out of 450), the ProM plugin for EPC soundness analysis 31 (out of 450) and YAWL Editor 15 (out of 203).

In contrast, our Prolog-based tool needed only 65 seconds for analyzing *all* models from the repository. This time included searches for 12 patterns that are related to control-flow errors as well as searches for more 12 patterns that are related to bad modeling style or suspiciously looking modeling elements.

After doing the analysis with the different tools, we compared the results. Because of subtle differences among the tools when it comes to defining the semantics of the OR-join [17,18], there were a few differences in the results of the tools. The analysis of these differences will be the subject of another paper; here it is sufficient to say that in cases of differences among the tools we looked at the model and selected the result that complied with the intuitive understanding of its semantics.

The comparison between our heuristic results and the exact results showed that the heuristics worked almost as good as the state-space exploring tools: For all models which have been categorized as not being sound by the "exact" tools, our Prolog program also found at least one pattern that (likely) shows a violation of the soundness property, i.e. we have had no "false negatives". On the other hand, our program warned about a possible soundness violation for exactly one model that turned out to be sound, i.e. we have had only one "false positive". It is worth mentioning that the model, for which this false positive occurred, was taken from [19] where it has been published as an example for bad modeling that should be improved.

YAWL Editor also allows checking whether an OR-join should be replaced (as this was the case for the model fragments in Fig. 3 (a)-(c)) which would not affect soundness, but can be considered as an improvement in the modeling style as discussed in Sect. 5. Our heuristics missed just one case where such a substitution was possible (false negative), but did not produce any incorrect warnings (false positives).

8   Related Work

Logic-based and pattern-matching approaches have been used in many published approaches for finding modeling errors. Their main application area is the detection of syntactical errors and inconsistencies within one model or between different models [20]. Our approach adds one more perspective by also detecting control-flow errors (like deadlocks) and even pragmatic issues.

Störrle [21] showed that a representation of models as logic facts can be very useful for querying model repositories as well.

ArgoUML [22] is an excellent example of a user-friendly modeling tool that runs tests in background in order to give the modeler feedback about possible improvements. So-called "design critics" inform the modeler about possible problems. The user is allowed to create own design critics. The most design critics currently implemented in ArgoUML either work on a rather syntactical level (i.e. they check consistency requirements and constraints that have to be followed according to the UML standard) or test whether modeling style conventions [23] are followed.

Work on error patterns for business process models has been done by different authors ([24,25,26,27]. As none of these pattern systems considered all three types of connectors that can occur in EPC models (OR, AND and XOR), they are of limited use for the assessment of business process models in languages in which all these connectors can be found.

The approach by Mendling [6] which applies reduction rules for finding errors considers all kinds of connectors. It is able to find a great part of errors in EPC models. Therefore, we used the reduction rules given in [6] as one starting point for creating our patterns. However, Mendling does not lay importance on the completeness of the rules; he uses reduction rules mainly for simplifying a model before using a state-space exploring algorithm for validating the model.

All the pattern systems mentioned above (which all share a set of basic common patterns) are reflected in our rules for detecting control-flow errors. Our definition of matching splits and joins, which is one of the most fundamental rules of our rule system, was inspired by the use of traps and handles in workflow nets for finding control-flow errors [28,5].

9 Conclusions and Directions for Further Research

In our analysis of a large number of business process models, we found that our pattern-based approach performed almost as good as tools that apply model-checking when it comes to detect soundness violations and OR-joins that should be replaced by XOR- or AND-joins. An advantage of our approach is that it produced a result very fast while the other tools suffered from the symptoms of state-space explosion and failed to deliver a result for some models. Furthermore, we have also patterns for hard-to-read parts of the model and "suspiciously looking" parts of the model that might indicate a pragmatic error even if the model is sound.

Using Prolog, the patterns can be specified very easily, and it is possible to add new patterns (for example for applying organization-wide style conventions) very quickly. However, our pattern-based approach does not necessarily have to be used with Prolog or another logic-based language. We have already implemented a pattern-finding algorithm in the open source Eclipse-based modeling tool *bflow*^{*2} using the languages *oAW Check* and *XTend* from the openArchitectureWare model management framework [29]. With this implementation, *bflow** gives the modeler immediate feedback about possible modeling problems.

Currently, we are working on an implementation using the query language BPMN-Q [30]. This will allow us to apply our approach to BPMN models.

One future direction of research is to consider more sophisticated modeling elements (like exceptions or cancellation) that exist in languages like BPMN or YAWL. This will allow us to deal with more complex patterns like the ones we have discussed in [9].

We are also researching problems that can be found by analyzing the textual description of events and functions. We are already able to find some problems this way. For example, if an OR-split is followed by both an event and its negation (as "article is (not) available)" in Fig. 1), it is very likely that the OR-split has to be replaced by an XOR-split, because both events cannot happen together.

References

1. Rutar, N., Almazan, C.B., Foster, J.S.: A comparison of bug finding tools for java. In: ISSRE, pp. 245–256 (2004)
2. van der Aalst, W.M.: Formalization and verification of event-driven process chains. Information & Software Technology 41, 639–650 (1999)
3. Gruhn, V., Laue, R.: Checking properties of business process models with logic programming. In: Augusto, J.C., Barjis, J., Ultes-Nitsche, U. (eds.) MSVVEIS, pp. 84–93. INSTICC Press (2007)
4. van der Aalst, W.M.P.: Verification of workflow nets. In: Azéma, P., Balbo, G. (eds.) ICATPN 1997. LNCS, vol. 1248, pp. 407–426. Springer, Heidelberg (1997)
5. van der Aalst, W.M.P.: Structural characterizations of sound workflow nets. Computing Science Reports/23 (1996)

---

^{2} http://www.bflow.org

6. Mendling, J.: Detection and Prediction of Errors in EPC Business Process Models. PhD thesis, Vienna University of Economics and Business Administration (2007)
7. Fillies, C., Weichhardt, F.: Towards the corporate semantic process web. In: Berliner XML Tage, pp. 78–90 (2003)
8. Thomas, O., Fellmann, M.: Semantic EPC: Enhancing process modeling using ontology languages. In: Hepp, M., Hinkelmann, K., Karagiannis, D., Klein, R., Stojanovic, N. (eds.) SBPM. CEUR Workshop Proceedings, vol. 251. CEUR-WS.org (2007)
9. Gruhn, V., Laue, R.: Good and bad excuses for unstructured business process models. In: Proceedings of 12th European Conference on Pattern Languages of Programs (EuroPLoP 2007) (2007)
10. van Dongen, B.F., van der Aalst, W.M.P., Verbeek, H.M.W.: Verification of EPCs: Using reduction rules and Petri nets. In: Pastor, Ó., Falcão e Cunha, J. (eds.) CAiSE 2005. LNCS, vol. 3520, pp. 372–386. Springer, Heidelberg (2005)
11. Cuntz, N., Kindler, E.: On the semantics of EPCs: Efficient calculation and simulation. In: EPK 2004: Geschäftsprozessmanagement mit Ereignisgesteuerten Prozessketten, Proceedings, pp. 7–26 (2004)
12. Cuntz, N., Freiheit, J., Kindler, E.: On the Semantics of EPCs: Faster calculation for EPCs with small state spaces. In: EPK 2005, Geschäftsprozessmanagement mit Ereignisgesteuerten Prozessketten, pp. 7–23 (2005)
13. Barborka, P., Helm, L., Köldorfer, G., Mendling, J., Neumann, G., van Dongen, B.F., Verbeek, E., van der Aalst, W.M.P.: Integration of EPC-related tools with ProM. In: Nüttgens, M., Rump, F.J., Mendling, J. (eds.) EPK. CEUR Workshop Proceedings, vol. 224, pp. 105–120. CEUR-WS.org (2006)
14. Wynn, M.T.: Semantics, Verification, and Implementation of Workflows with Cancellation Regions and OR-joins. PhD thesis, Queensland University of Technology Brisbane, Australia (2006)
15. Wynn, M.T., Verbeek, H., van der Aalst, W.M.P., Edmond, D.: Business process verification - finally a reality! Business Process Management Journal 15, 74–92 (2009)
16. Decker, G., Mendling, J.: Instantiation semantics for process models. In: Proceedings of the 6th International Conference on Business Process Management, Milan, Italy (2008)
17. van der Aalst, W.M.P., Desel, J., Kindler, E.: On the semantics of EPCs: A vicious circle. In: EPK 2004, Geschäftsprozessmanagement mit Ereignisgesteuerten Prozessketten, pp. 71–79 (2002)
18. Dumas, M., Grosskopf, A., Hettel, T., Wynn, M.: Semantics of BPMN process models with or-joins. Technical Report Preprint 7261, Queensland University of Technology, Brisbane (2007)
19. Mendling, J., Reijers, H.A., van der Aalst, W.M.P.: Seven process modeling guidelines (7pmg). Technical Report QUT ePrints, Report 12340, Queensland University of Technology (2008)
20. Finkelstein, A.C.W., Gabbay, D., Hunter, A., Kramer, J., Nuseibeh, B.: Inconsistency handling in multiperspective specifications. IEEE Trans. Softw. Eng. 20, 569–578 (1994)
21. Störrle, H.: A prolog-based approach to representing and querying software engineering models. In: Cox, P.T., Fish, A., Howse, J. (eds.) VLL. CEUR Workshop Proceedings, vol. 274, pp. 71–83. CEUR-WS.org (2007)
22. Robbins, J.E., Redmiles, D.F.: Cognitive support, UML adherence, and XMI interchange in Argo/UML. Information & Software Technology 42, 79–89 (2000)

23. Ambler, S.W.: The Elements of UML Style. Cambridge University Press, Cambridge (2003)
24. Onoda, S., Ikkai, Y., Kobayashi, T., Komoda, N.: Definition of deadlock patterns for business processes workflow models. In: Proceedings of the 32nd Annual Hawaii International Conference on System Sciences, vol. 5, p. 5065. IEEE Computer Society, Los Alamitos (1999)
25. Koehler, J., Vanhatalo, J.: Process anti-patterns: How to avoid the common traps of business process modeling, part 1 - modelling control flow. IBM WebSphere Developer Technical Journal (2007)
26. Liu, R., Kumar, A.: An analysis and taxonomy of unstructured workflows. In: Business Process Management, pp. 268–284 (2005)
27. Smith, G.: Improving process model quality to drive BPM project success (2008), http://www.bpm.com/improving-process-model-quality-to-drive-bpm-project-success.html (accessed November 1, 2008)
28. van der Aalst, W.M.P.: The Application of Petri Nets to Workflow Management. The Journal of Circuits, Systems and Computers 8(1), 21–66 (1998)
29. Kühne, S., Kern, H., Gruhn, V., Laue, R.: Business process modelling with continuous validation. In: Pautasso, C., Koehler, J. (eds.) MDE4BPM 2008 – 1st International Workshop on Model-Driven Engineering for Business Process Management (2008)
30. Awad, A.: BPMN-Q: A language to query business processes. In: Reichert, M., Strecker, S., Turowski, K. (eds.) EMISA, GI. LNI, vol. P-119, pp. 115–128 (2007)

Simulating Liquidity in Value and Supply Chains

Wim Laurier and Geert Poels

Department of Management Information and Operational Management,
Faculty of Economics and Business Administration, Ghent University, Tweekerkenstraat 2,
9000 Ghent, Belgium
{wim.laurier,geert.poels}@ugent.be

Abstract. This paper provides an ontology-based set of Petri-nets for simulating the effect of business process changes on an organisation's liquidity, and demonstrates that certain types of business process redesign can increase or reduce the amount of external funding that is required to prevent an organisation from defaulting on its debt. This debt defaulting may lead to proliferating liquidity constraints for subsequent supply chain partners. Consequently, this paper provides a proper toolkit for assessing and mitigating the propagation of liquidity constraints in supply chains. The paper uses the accounting-based Resource-Event-Agent ontology to create workflow patterns for modelling exchanges between supply chain partners and for the value chains that represent an organisation's internal processes. Both the exchange and internal processes continuously convert money into resources and vice versa. These models for money to resource and resource to money conversions are then used for constructing supply chain models for liquidity modelling and analysis.

Keywords: Resource-Event-Agent Ontology, Petri-net, Simulation, Business Process Management, Workflow Model, Business Performance, Liquidity.

1 Introduction

In recent years, business process and workflow management have concentrated on modelling, simulating and evaluating the physical and informational aspects of production and trade processes. [1-5] These processes are embedded in the enterprise's value creating activities that support the purpose of being profitable in the long term. This profitability is essentially a financial measure that is usually approximated in operational environments by efficiency and effectiveness measures such as cycle time, process time and production per time unit. Although these measures are generally believed to be connected to the profitability measure, the financial effect of changes in those operational measures is hard to simulate. What is needed, especially in a credit crunch during which cost thinking and consequently cost cutting soars, is a framework that allows us to simulate the effect of operational changes to an organisation's cash position and consequently its need for external funding (e.g. loans). This cash position is essential for an enterprise's continuity, and consequently its ability to reach its long-term goals, as insufficient cash supplies hamper the acquisition of

A. Albani, J. Barjis, and J.L.G. Dietz (Eds.): CIAO!/EOMAS 2009, LNBIP 34, pp. 40–54, 2009.

inputs for the organisation's production processes and consequently also its ability to generate revenue from production process outputs.

In the systems design community, also the need *"to separate the stable ontological essence of an enterprise from the variable way in which it is realized and implemented"* [6] has been recognized, which has lead to the creation of various enterprise and business ontologies (e.g. enterprise ontology[6], e3-value[7], REA[8], BMO[9]) These ontologies describe the economic reality as a collection of constructs and axioms. The former create a controlled vocabulary for describing the business reality, whereas the latter represent *"fundamental truths that are always observed to be valid and for which there are no counterexamples or exceptions"* [10].

As this paper intends to address the financial consequences of operational decisions within a stable framework that is sufficiently generic to be applied to different enterprises, the REA ontology was selected as the ontological basis for the presented simulation framework. The REA ontology was originally designed for sharing data concerning economic phenomena between accountants and non-accountants [11]. It has been applied in several other sub-domains of business (e.g. value chain [12, 13] and supply chain modelling [14, 15]). In this paper, the REA ontology provides the conceptual basis for a set of Petri-net model construction patterns, which can be used for constructing business process simulation models. As Petri-nets provide an intuitive mathematically based formalisation syntax for process representation that is widely used and supported by many tools for business process simulation they provide the preferred syntax for the models in this paper. Committing these simulation models to REA not only allows checking their syntactic correctness, but also their semantic correctness (i.e. Do they represent proper business logic?). Syntax checking and simple simulation facilities are provided by the WoPeD[1] tool, which is used to create the exhibits in this paper, whereas more complex model simulations can be performed by the CPN tool[2].

Using the proposed set of REA-based Petri-net workflow model construction patterns, the paper proceeds by presenting generic workflow models for exchanges between supply chain partners and for the internal value chains that ensure an organization's continuity and sustainability. These exchange and value chain models are next combined into generic elements for constructing supply chain models that comply with the REA axiomatisation. The paper then demonstrates that these supply chain models can be used to identify an organisation's dependency on external funding and to analyze the effect of business process changes on the organization's liquidity (or cash position) and its need for external funding.

The Petri-net workflow model construction patterns are presented in section 2, together with the basics of the REA ontology. Sections 3 and 4 present the generic workflow models for exchanges and internal value chains. Then section 5 presents the integration of the latter two types of models into supply chain models and explains how these models can be analyzed to identify an organization's need for external funding. Section 6 presents an illustrative case in which our proposed modelling framework is applied to analyze the financial consequences of an intelligent workflow design in the invoicing process. Section 7 presents conclusions and future research.

---

[1] www.woped.org
[2] http://wiki.daimi.au.dk/cpntools/cpntools.wiki

2 Atomic REA Model Construction Patterns

This section presents three workflow model patterns in a Petri-net formalisation that commit to the REA ontology and its axiomatisation [16]. Committing to these axioms ensures modellers that the economic rationale is represented in their models. This rationale is what is generally abstracted from in conventional business process and workflow models, as the financial resources that are generated by these business processes and required for acquiring the inputs to these processes are often not shown. This omission helps designers to focus on the essential parts of the business process design but prevents them from assessing the financial consequences of process redesigns. These financial consequences of particular business process designs are the topic of this paper, as liquidity is a consequence of revenue generated from business process outputs and a prerequisite for acquiring the inputs to these processes.

The controlled REA vocabulary specifies four main concepts (i.e. resources, events, agents and economic units) [11]. Resources (e.g. goods and services) represent objects that are scarce, have utility and are under the control of economic units (i.e. legal or natural persons) that can use or dispose of them during events [17, 18]. These events are initiated and controlled by agents (i.e. natural persons). As both economic units and agents represent persons, a more precise distinction needs to be made. This distinction specifies that agents (e.g. employee) act on behalf of economic units (e.g. employer, enterprise) [14].

The REA ontology has three basic axioms:

- REA axiom 1: *"At least one inflow event and one outflow event exist for each economic resource; conversely inflow and outflow events must affect identifiable resources."*
- REA axiom 2: *"All events effecting an outflow must be eventually paired in duality relationships with events effecting an inflow and vice-versa."*
- REA axiom 3: *"Each exchange needs an instance of both the inside and outside subsets."*

The first REA axiom specifies that each resource has an origin (i.e. inflow) and a purpose (i.e. outflow) in a production process (i.e. inputs and outputs) or an exchange (i.e. sales and purchases). Additionally, the first REA axiom requires economic events to affect resources. The second REA axiom defines the duality principle that materializes as claims (e.g. accounts receivable), which model decrements[3] (e.g. a shipment) that must be succeeded by increments[4] (e.g. receiving a payment), and as liabilities (e.g. accounts payable), which model increments (e.g. goods receipt) that must be succeeded by decrements (e.g. payment). The third REA axiom applies strictly to exchanges, stipulating that an exchange requires at least two parties (i.e. one taking the inside view (i.e. the enterprise that sells or buys) and one or more that are considered outside trading partners (e.g. customer, supplier) for this inside party).

As the models in the remainder of this paper are articulated in a workflow Petri-net formalisation annotated with words from the REA ontology vocabulary, some essential differences between the REA and the workflow idiom need to be addressed. The

---

[3] i.e. losing control over resources.
[4] i.e. gaining control over resources.

most essential difference between the REA and workflow idiom is the use of the word 'resource'. In workflow models, the REA agents, in their role of 'performers of activities' are considered as resources because workflow resources represent both animate and inanimate means of production [5]. In the REA idiom, resources represent solely inanimate means of production such as goods, services and money. We will stick to the REA interpretation in the remainder of this paper and thus clearly distinguish animate (i.e. agents) and inanimate (i.e. resources) means of production. Another difference between the REA and workflow vocabulary is the use of the word event. Events that are atomic from an economic point of view may represent entire operational processes in workflow models. Consequently, events as used in the REA vocabulary may represent complete workflow models with numerous activities or tasks that all contribute to the realisation of the event.

The event template (fig. 1) shows the structure in which conventional workflow models need to be embedded to allow modelling the flows of value (i.e. a measure for the amount of money a resource can potentially generate) and money throughout supply and value chains. The double-edged box indicates where workflow models for value creating activities, such as logistic and production processes need to be inserted. According to the REA conceptualisation, agents, who represent persons that act on behalf of economic units, participate in economic events [14]. The economic units represent structures that can acquire, use and dispose of resources (e.g. organisations, enterprises, households). The decrement box (i.e. transition) shows that each (transfer or transformation) process requires inputs from the viewpoint of the economic unit that loses the control over these resources, whereas the increment box represents the outputs of such a process from the viewpoint of the economic unit that gains control over the converted or traded resources or resource bundles. The two economic units involved are different parties in case of trade and one and the same in case of conversions. The event template can be combined with itself to create entire supply chain models since every sink (i.e. right-hand side resource) of an event template equals the source (i.e. left-hand side resource) of a subsequent event template.

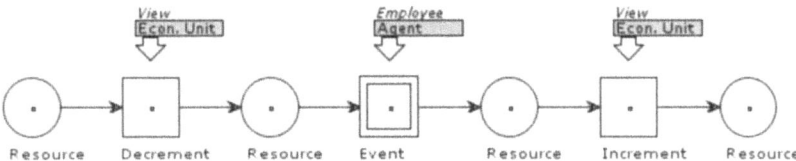

Fig. 1. Event template

The duality templates (fig. 2a & 2b) represent the creation of claims (fig. 2a), which need to be settled by gaining control over new resources when economic units lose control over other resources, and liabilities (fig. 2b), which need to be settled by losing control over resources when economic units gain control over new resources. These dualities link oppositely directed resources flows. Conventionally, these resource flows represent goods or service flows that are paired with money flows. However, money flows can be paired with other money flows (e.g. loans) and goods and service flows can be paired with other goods and service flows (i.e. barter trade).

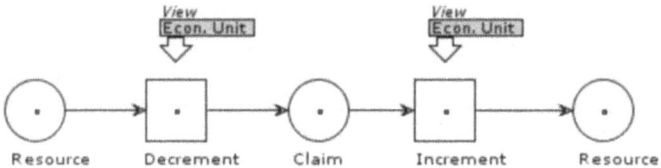

Fig. 2a. Claim Duality template

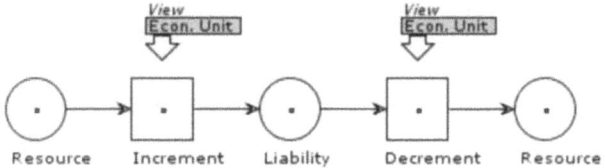

Fig. 2b. Liability Duality template

In contrast to the event template, the increments and decrements in a duality template need to be related to the same economic unit as they represent this economic unit's acquisition and revenue cycles [13] that model how economic units acquire and dispose of resources in exchanges.

3   Exchange Configurations

In this section, the potential configurations for exchanges are represented using the model construction patterns of the previous section. The potential for configuration is limited to two pattern variations as the claims and liabilities in exchange models need to mirror each other (i.e. a claim for a trade partner implies a liability for another trade partner and vice versa). The claims or liabilities that represent the revenue cycle of a seller are coupled with the claims or liabilities that represent the acquisition cycle of a buyer via transfer events that represent either the transfer of products (i.e. goods and services) or the transfer of money. These transfer events are executed by agents that are accountable for them (e.g. a carrier).

Fig. 3a shows a conventional situation where sellers temporary pre-finance their sales, creating a claim (i.e. accounts receivable) that is a liability (i.e. accounts payable) for the buyer. Consequently the seller is exposed to default risk and requires the financial means (e.g. equity or long term capital) to fund these unpaid sales. The workflow model also reveals a logical sequence between the product and the money transfer (i.e. the product transfer precedes the money transfer). Consequently, the lifespan of the claim is determined by the duration of both transfers and the lifespan of the liability. Hence, reducing the term of payment and the duration of the transfers means reducing the need for external funding as the number of claims that needs to be financed is reduced.

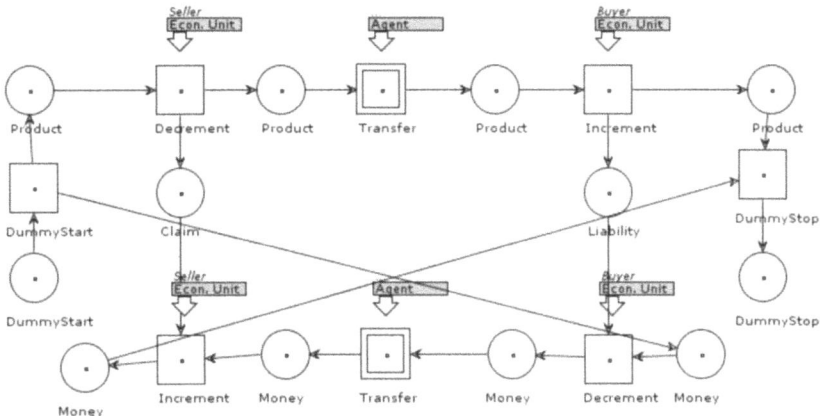

Fig. 3a. Buyer liable

Fig. 3b shows the opposite of fig. 3a, meaning that in this exchange process model, buyers pre-finance their future purchases. As a result, sellers are liable to the buyers creating a liability for the seller (i.e. payments received on account of orders) and a claim for the buyer (i.e. orders in progress). This practice is rather exceptional, although it is applied in long-term projects (e.g. property development). In this situation, the buyer is exposed to default risk and requires sufficient financial means to fund these undelivered purchases. The according workflow model specifies that the money transfer precedes the product transfer, which makes the lifespan of the claim a sum of the duration of both transfer events and the lifespan of the liability. As in the previous workflow model, the duration of the claim, and consequently the amount of required funding, can be reduced speeding up the transfers and reducing the lifespan of the liability.

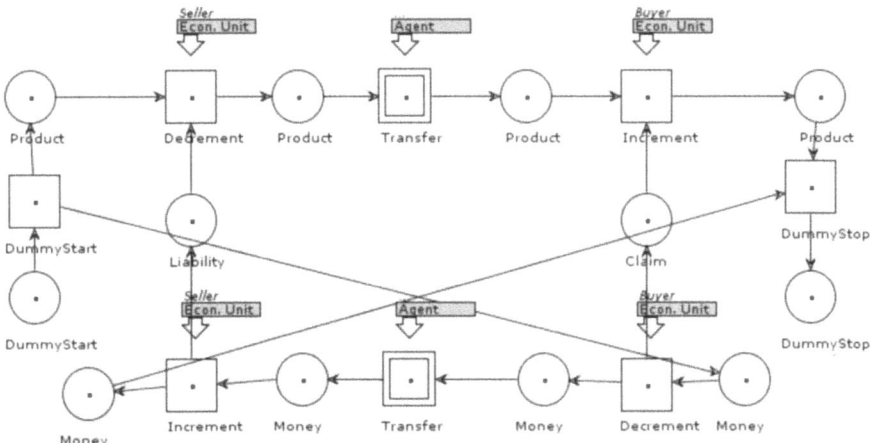

Fig. 3b. Seller liable

4   Value Chain Configurations

This section presents the four potential configurations of an economic unit's value chain, which contains an acquisition, a conversion, a revenue and a financing cycle [13]. The acquisition cycle exchanges money for products, whereas the revenue cycle exchanges products for money. The conversion cycles consume products (i.e. raw materials) producing new products, while the money conversions in the financing cycle model the process of approving payments, which converts incoming money flows into outgoing money flows. Such money conversions include converting incoming money flows from loans and equity to outgoing money flows for acquiring resources. They also include converting incoming revenue to outgoing instalments, dividends and money for resource acquisitions.

Both the acquisition and revenue cycles can be represented by claims or liabilities. In contrast to the exchange configurations in the preceding section, claims and liabilities do not need to mirror each other in value chain configurations since there are no opposing economic interests between the acquisition and the revenue cycle of an organisation. The acquisition and revenue cycle of the same economic unit both try to maximize its profit, whereas the acquisition and revenue cycles of trading partners try to maximize the profit of their respective economic unit. As a result, the value chain pattern has four variations. As agents participate in the transfer events, they also participate in the conversion events in the value chain patterns (e.g. employees). For the reason of conciseness, the models in this section abstract from the increment and decrement sides of conversion events, which are hidden inside the double-edged conversion box. Consequently, the models represent a trading-partner view on business, which abstracts from the increments and decrements in an enterprise's internal processes.

In fig. 4a, the organisation's acquisition and revenue cycles create liabilities to the organisation's suppliers and customers respectively. For organisations, this situation is ideal as customers pre-finance their own purchases and the suppliers finance their own sales. Consequently, organisations without long-term capital and equity could exist in theory. The workflow model reveals that the product (i.e. production process) and money (i.e. approving of payments) conversions are fully parallel but synchronized, which means that the lifespan of the liabilities is not directly determined by a

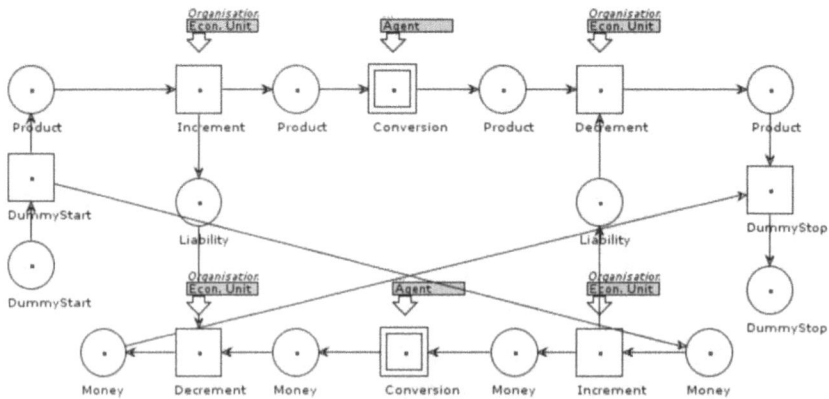

Fig. 4a. Economic Unit liable to Suppliers and Customers

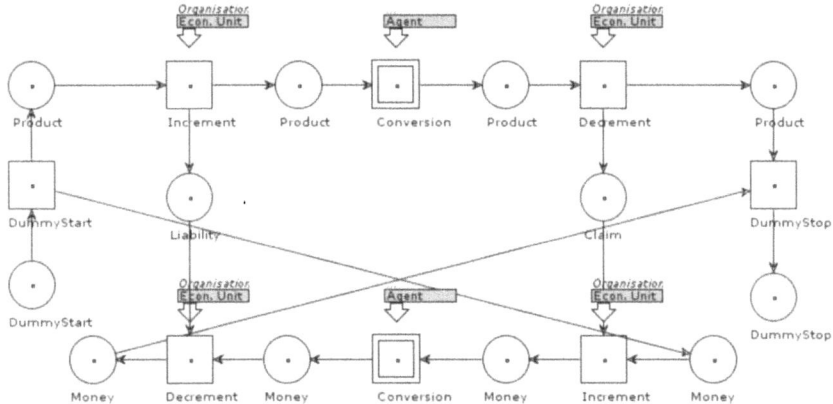

Fig. 4b. Economic Unit liable to Suppliers, claim on Customers

conversion although the duration of the synchronisation process (i.e. the entire work-flow) is determined by the longest of the conversions.

Fig. 4b, on the other hand, models a more conventional situation where suppliers pre-finance their sales (i.e. buyer liable). Consequently, the economic unit under re-view is liable to its supplier and has a claim on its customer. As a result, purchases provide (temporary) funding and sales require (temporary) funding. The workflow model for this value chain configuration reveals a logical sequence between the con-versions (i.e. the product conversion needs to precede the money conversion). This sequence stipulates that the lifespan of the liability to the suppliers is determined by the duration of both conversion processes (i.e. production process and approving payments) and the lifespan of the claim on the customer.

Fig. 4c models the converse of fig. 4b, representing customers that pre-finance their purchases (i.e. seller liable). As a result, the economic unit under review is liable to its customers and has a claim on its suppliers. Accordingly, sales provide tempo-rary funding and purchases require funding. The workflow model for this value chain configuration shows a sequence of events which specifies that approving payments needs to precede the creation of value in production processes. Consequently, the lifespan of the liability is determined by the duration of the conversion events and the lifespan of the claim, as was also the case for the preceding value chain configuration.

Finally, fig. 4d shows the opposite of fig. 4a, meaning that the economic unit under review pre-finances both its sales and purchases. Hence, both its suppliers and cus-tomers are liable to the economic unit under review. Consequently, both sales and purchases require external funding. This feature is also reflected in the workflow model as this workflow model suffers dead transitions [5], reflecting that this value chain configuration cannot be self-sustaining, which matches the requirement for external funding. This external funding would create additional money inflows that create liabilities owed to investors. In the workflow model this would mean breaking the loop into two mirrored value chain configuration with one claim and one liability each (i.e. fig. 4b & 4c).

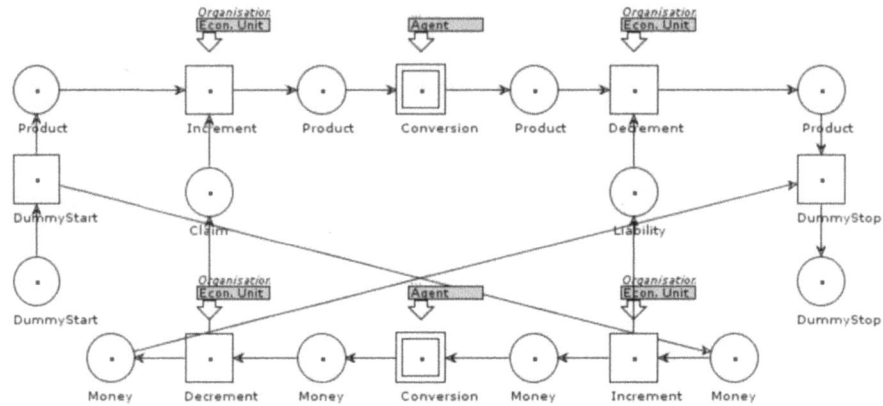

Fig. 4c. Economic Unit claim on Suppliers, liable to Customers

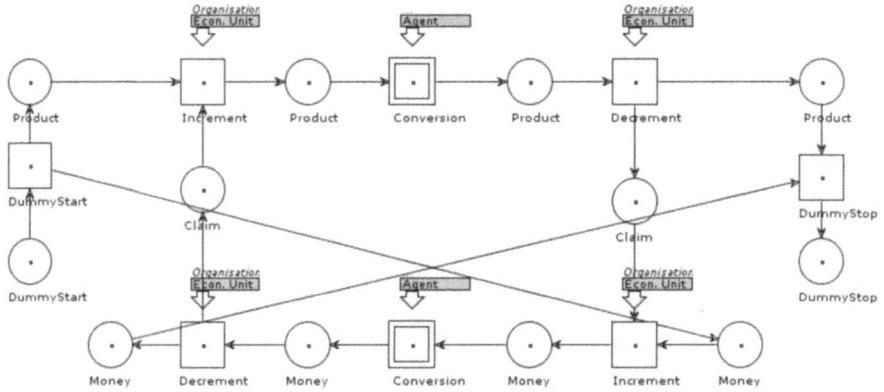

Fig. 4d. Economic Unit claim on Suppliers and Customers

5 Supply Chain Pattern

In this section, the value chain and exchange configuration patterns are combined to create supply chain links that symbolize an economic unit and its interface (i.e. exchange) with a subsequent supply chain partner. In prior sections, no constraints have been imposed on the lifespan of dualities (i.e. claims and liabilities). Consequently, no distinction could be made between claims and liabilities that are ought to exist for a few seconds (e.g. in a cash sale) and claims and liabilities that should exist for several weeks, months or years (e.g. invoices, loans). To resolve this, the exchange pattern (fig. 3a & 3b) has been extended with a business process that determines when liabilities are due (fig. 5). Such a business process is an example of a support activity in Porter's value chain [19]. Such an activity supports the economic processes in the economic events (fig. 1), which are containers for primary activities in Porter's value chain. According to the REA ontology, these primary activities are economic events

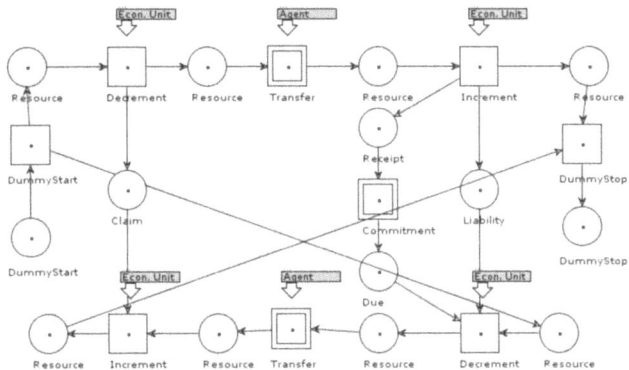

Fig. 5. Duality Lifecycle template

and support activities consist of business events, which do not imply losing or gaining ownership over resources but are notwithstanding worth monitoring and controlling [14]. The according workflow model (fig. 5) shows a business process (i.e. commitment) that determines the lifespan of the liability. This business process can be automated or executed by human agents. The process is coupled to the liability duality as valid liabilities can only be created from receipts and liabilities can only be due when all formalities have been completed (e.g. a payment for a purchase can only occur when the buyer received an invoice and the invoice is due). The template in fig. 5 is applicable to both exchange pattern variations in section 3; therefore it abstracts from 'money' and 'product' flows.

The first supply chain link (fig. 6a) creates a structure in which organisations are liable to their customers who pre-finance their purchases. This pattern variation combines fig. 4c with 3b, merging two rather exceptional configurations. The workflow model shows that the product delivery (i.e. product transfer) can be initiated (i.e. is 'enabled') as soon as the organisation received the payment (i.e. money transfer) and the delivery is due. Next to the expected lifespan of the buyer's claim (i.e. the duration of both transfers and the claiming process), the workflow model also indicates the maximal lifespan (i.e. the duration of both transfers, both conversions and the supplier's claim on its supplier). The minimal lifespan is achieved when there is a stock of already converted products, which allow us to reduce the lifespan of the liability to zero. The maximal lifespan occurs when the inputs to the conversions still need to be acquired. As both the right- and left-hand side of fig. 5a represent acquisition cycles that are implemented with claims, this supply chain pattern variation can be combined with itself to create entire pre-financed supply chains.

Whereas fig. 6a shows a supply chain link in which an organisation is liable to its customers, fig. 6b models the more conventional situation where organisations have claims on their customers and the lifespan of the according liabilities is determined by the invoicing process. This pattern variation combines fig. 4b with 3a. The workflow model reveals that product transfers can occur autonomously but money transfers occur only as a counter-value for product transfers when the according invoices are due. In this template, the lifespan of the liabilities is constrained by the invoicing process (i.e. the minimal and expected lifespan) and the conversions and transfers

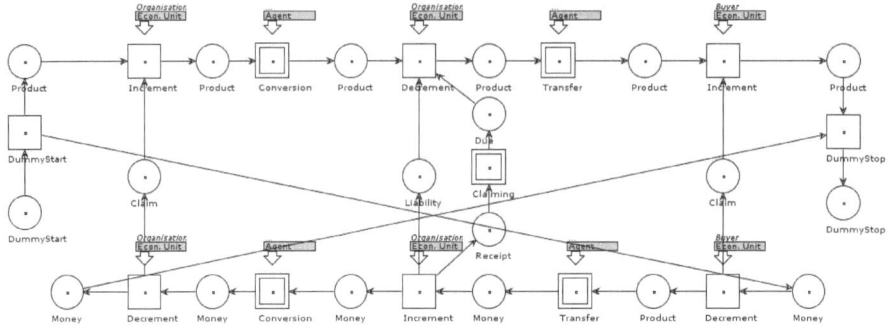

Fig. 6a. Supply Chain link, claim on Suppliers and liable to Customers

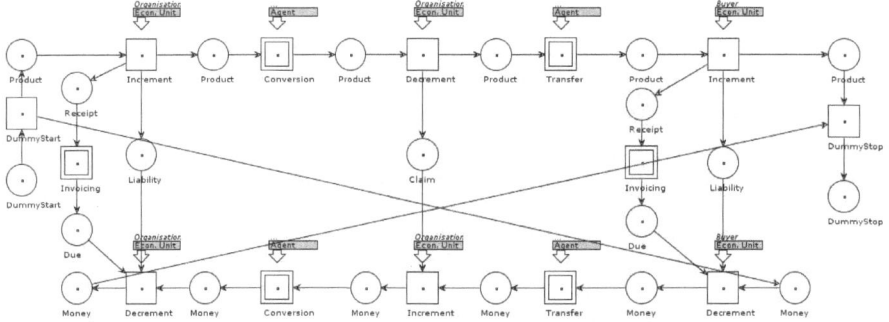

Fig. 6b. Supply Chain link, claim on Customers and liable to Suppliers

further up (i.e. right-hand side) the supply chain. The maximal lifespan of the liability on the left-hand side of fig. 6b, is determined by the duration of both transfers and conversions and by the lifespan of the liability on the right-hand side of fig. 6b. As with the preceding link (fig. 6a), fig. 6b can be combined with itself to create entire supply chain models in which suppliers temporarily finance their sales.

Similar and more complex supply chains can be constructed applying the duality lifespan template to the variations of the exchange pattern shown in section 3, combining them with the value chain patterns in section 4.

6   Discussion

This section demonstrates and illustrates the effect of the duration of business processes on the amount of external funding required. This effect reveals the economic rationale for business process redesign [20] and management [21] as it is currently practiced. As an example, we take the conventional supply chain template where organisations are only liable to organisations upstream the supply chain (i.e. their suppliers) (fig. 6b) and discuss the implications of different choices regarding business process configuration to the required amount of external funding. The presented theoretical example addresses an archetypal supermarket first and an archetypal retail store next. Both types of organisation perform similar activities but the underlying business processes may be configured differently.

We assume that, due to its market power, the supermarket can obtain a 3 month (i.e. 90 days) pre-financing period from its suppliers. This pre-financing period is to be represented as a time trigger in a timed Petri-net representation of the invoicing process of the supermarket's supplier. This invoicing process is represented at the left-hand side of fig. 7. The product conversion process of the supermarket is a complex logistic process in which products reside 2 months (i.e. 60 days) on average. Next, they have an average shelf-life of approximately one week (i.e. 5 days), which is represented as the product transfer process in fig. 7. Considering that most supermarket sales are cash sales, the duration of the right-hand side invoicing process is reduced to 0 days. Subsequently, it takes 2 days on average to put the cash received selling product on the supermarket's bank account, which is represented as the money transfer process. Once the money is on the bank account it can be spent again. However, it takes another week (i.e. 5 days) on average to approve payments (i.e. one meeting every two weeks). Consequently, abstracting from the profit margin, it takes the supermarket 72 days (i.e. 60 + 5 + 0 + 2 + 5) on average to have its money available for a subsequent business cycle. As the supermarket obtained a 90 day pre-financing period, it has an additional 18 days of funding it can use to fund other things. This means that the supermarket can use 5% (i.e. 18 days / 90 days * 3 months / 12 months) of its annual turnover to fund fixed assets.

The retail store, on the other hand, has no market power and can only obtain a 30 day pre-financing period from its suppliers. It takes the retailer one week (i.e. 5 days) on average to take the purchased products from its warehouse to the shelf. In contrast to the supermarket, the shelf-life of retail products is much longer e.g. 1½ month (i.e. 50 days). As with the supermarket, most retail sales are cash sales which take 0 days to complete. Further, it takes 2 days on average to transfer the cash from the counter to the bank account. Then the retailer checks its accounts payable once a week (i.e. the money conversion takes 3 days on average). This results in a business process cycle that takes 60 days on average (i.e. 5 + 50 + 0 + 2 + 3) which is twice the period that has been pre-financed by the supplier. Consequently, the retailer continuously needs to fund one month of turnover with other financial means (e.g. loans, equity).

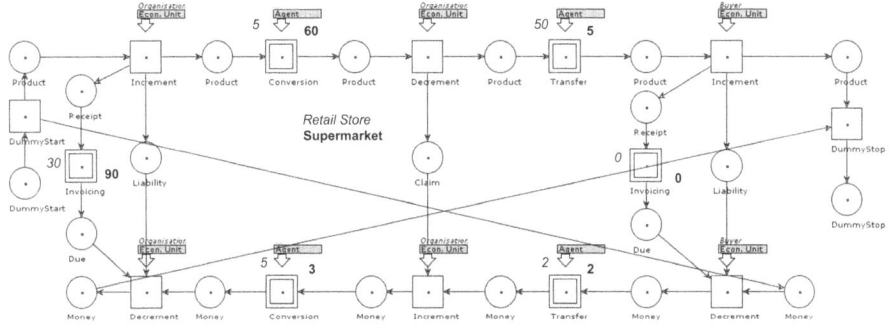

Fig. 7. Exhibit: Supermarket and Retail Store

Both the retailer and the supermarket can enhance their liquidity (i.e. reducing the need for extra funding) by reducing the duration of the conversion and transfer processes. The duration of their invoicing process, which was set 0 in both examples, also indicates a potential risk for liquidity constraints further down the supply chain (i.e. customers and customer's customers) for economic units further up the supply chain (i.e. suppliers and supplier's suppliers) as liquidity constraints deadlock the money transfers that enable economic units to close their internal value chain. This (temporal) deadlock increases the need for external funding or imposes liquidity constraints if additional money is not found. Other potential solutions are shortening the pre-financing period for customers (which may induce new liquidity constraints for the customers) demanding longer pre-financing periods from suppliers (which may create liquidity constraints for them) or shortening the conversion and transfer processes in the business model, which is a motivation for business process redesign.

Deadlock can also appear for the supply chain template in fig. 6a when the supplier does not deliver on time, as the customer cannot generate new financial means from products that it can sell to its own customers. The difference with the template in fig. 6b is that the liquidity constraints move further down (i.e. towards customers and customer's customers) the supply chain instead of further up the chain. For the symmetric value chain templates (fig. 4a & 4d) the situation is also rather straightforward. For the double liability template (fig. 4a) the liquidity is determined by the equilibrium between the opposing processes (i.e. product and money conversions) and their ability to generate resources. For the double claim (fig. 4d) the situation is even simpler as this template is not able to generate its own means and needs to be funded completely.

7   Conclusions and Future Research

This paper presented an ontology-based modelling framework for evaluating the financial consequences of alternative business process design choices. This framework is especially useful for analysing the amount of external funding required. First, the main constructs and axioms of the REA ontology, which provides a conceptual basis for the modelling framework, were introduced. Next, these constructs and axioms were used to create workflow model construction patterns (using a Petri-net formalisation). These patterns were subsequently used to build modelling templates for the value creating processes of economic units and for exchanges between supply chain partners. The paper presented two alternative configurations for exchanges and four alternative configurations for value creating business processes. These modelling template variations can be combined to create supply chain modelling templates, two of which (i.e. the more typical cases) were shown in the paper.

For each of these exchange, value chain and supply chain configurations, the corresponding Petri-net modelling template was analysed to assess its ability to generate its own financial resources. These analyses showed that requirements for external funding depend upon the lifespan of claims and liabilities and the duration of operational processes (conversions and transfers). Based on the analysis, suggestions were made to reduce the dependency on external funding. These suggestions can be implemented through the redesign and management of business processes. Among these redesign efforts, especially the redesigns that reduce value chain cycle time (i.e. the time between the acquisition of a resource and the moment a new similar resource can be acquired with the means generated by the former resource) were identified as useful for reducing an organisation's dependency upon external funding since they reduce the amount of assets that is to be funded with liabilities (e.g. debt).

In the future, we intend to apply the templates presented here to detailed business process designs instead of the highly abstract aggregate process models used in this paper. Consequently, claims and liability structures will be identified for each individual resource sold (e.g. product and services) or purchased (e.g. raw materials, labour, fixed assets). This will allow us to assess the consequences of individual business process redesigns to an organisation's liquidity. These model simulations will involve modelling timed Petri-nets and simulating their behaviour using the advanced simulation and monitoring capabilities provided by the CPN tool. In a later stage, these simulation efforts could evolve towards the development of a business process management software tool that supports managers in identifying and motivating priorities among potential business process redesigns by showing the financial impact of these redesigns.

References

1. Davis, R., Brabänder, E.: ARIS design platform: getting started with BPM. Springer, New York (2007)
2. Stefanovic, D., Stefanovic, N., Radenkovic, B.: Supply network modelling and simulation methodology. Simulation Modelling Practice and Theory 17, 743–766 (2009)
3. Cho, H., Kulvatunyou, B., Jeong, H., Jones, A.: Using business process specifications and agents to integrate a scenario-driven supply chain. International Journal of Computer Integrated Manufacturing 17, 546–560 (2004)
4. Vergidis, K., Tiwari, A., Majeed, B.: Business Process Analysis and Optimization: Beyond Reengineering. IEEE Transactions on Systems, Man, and Cybernetics, Part C: Applications and Reviews 38, 69–82 (2008)
5. van der Aalst, W.M.P., Hee, K.M.v.: Workflow management: models, methods and systems. MIT Press, Cambridge (2002)
6. Dietz, J.L.G.: Enterprise ontology: theory and methodology. Springer, Berlin (2006)
7. Gordijn, J.: Value-based requirements Engineering: Exploring innovative e-commerce ideas. Exact Sciences, Phd. Free University of Amsterdam, Amsterdam (2002)
8. Geerts, G.L., McCarthy, W.E.: An ontological analysis of the economic primitives of the extended-REA enterprise information architecture. International Journal of Accounting Information Systems 3, 1–16 (2002)
9. Osterwalder, A.: The Business Model Ontology - a proposition in a design science approach. Ecole des Hautes Etudes Commerciales, University of Lausanne, Lausanne (2004)
10. Bahrami, A.: Object oriented systems development. Irwin/McGraw-Hill, Boston, Mass (1999)
11. McCarthy, W.E.: The REA Accounting Model: A Generalized Framework for Accounting Systems in a Shared Data Environment. Accounting Review 57, 554 (1982)
12. Hruby, P.: Model-driven design using business patterns. Springer, Berlin (2006)
13. Dunn, C.L., Cherrington, J.O., Hollander, A.S.: Enterprise information systems: a pattern-based approach. McGraw-Hill/Irwin, Boston (2005)

14. ISO/IEC: Information technology - Business Operational View Part 4: Business transaction scenario - Accounting and economic ontology. ISO/IEC FDIS 15944-4: 2007(E) (2007)
15. Haugen, R., McCarthy, W.E.: REA, a semantic model for Internet supply chain collaboration. Business Object Component Workshop VI: Enterprise Application Integration (OOPSLA 2000) (2000)
16. Geerts, G.L., McCarthy, W.E.: The Ontological Foundation of REA Enterprise Information Systems. Michigan State University (2004)
17. Ijiri, Y.: Theory of accounting measurement. American Accounting Association, Sarasota, Fla (1975)
18. Yu, S.C.: The structure of accounting theory. University Presses of Florida, Gainesville (1976)
19. Porter, M.E., Millar, V.E.: How information gives you competitive advantage. Harvard Business Review 63, 149–160 (1985)
20. Reijers, H.A., Liman Mansar, S.: Best practices in business process redesign: an overview and qualitative evaluation of successful redesign heuristics. Omega 33, 283–306 (2005)
21. Schonberger, R.: Best practices in lean six sigma process improvement: a deeper look. John Wiley & Sons, Hoboken (2008)

Complexity Levels of Representing Dynamics in EA Planning

Stephan Aier, Bettina Gleichauf, Jan Saat, and Robert Winter

Institute of Information Management
University of St. Gallen
Müller-Friedberg-Strasse 8
CH-9000 St. Gallen
{stephan.aier,bettina.gleichauf,jan.saat,robert.winter}@unisg.ch

Abstract. Enterprise Architecture (EA) models provide information on the fundamental as-is structure of a company or governmental agency and thus serve as an informational basis for informed decisions in enterprise transformation projects. At the same time EA models provide a means to develop and visualize to-be states in the EA planning process. Results of a literature review and implications from industry practices show that existing EA planning processes do not sufficiently cover dynamic aspects in EA planning. This paper conceptualizes seven levels of complexity for structuring EA planning dynamics by a system of interrelated as-is and to-be models. While level 1 represents the lowest complexity with non-connected as-is and to-be models, level 7 covers a multi-period planning process also taking plan deviations during transformation phases into account. Based on these complexity levels, a multi-stage evolution of EA planning processes is proposed which develops non-dynamic as-is EA modeling into full-scale EA planning.

Keywords: EA planning, EA modeling, dynamics of EA.

1 Introduction

The ANSI/IEEE Standard 1471-2000 defines architecture as "the fundamental organization of a system, embodied in its components, their relationships to each other and the environment, and the principles governing its design and evolution" [1]. Most authors agree that enterprise architecture (EA) targets a *holistic scope* and therefore provides a broad and aggregate view of an entire corporation or government agency [2, 3] covering strategic aspects, organizational structure, business processes, software and data, as well as IT infrastructure [4, 5, 6]. Enterprise architecture management can provide systematic support to organizational change that affects business structures as well as IT structures by providing constructional principles for designing the enterprise [7]. In order to provide support for transformation in an efficient way, EA has to be driven by business and/or IT oriented application scenarios [8] based on stakeholders concerns [9, 10, 11] (*goal orientation*) [3, 6]. Since the involvement of heterogeneous stakeholder groups may create conflicting requirements in a complex

A. Albani, J. Barjis, and J.L.G. Dietz (Eds.): CIAO!/EOMAS 2009, LNBIP 34, pp. 55–69, 2009.

environment, an appropriate documentation and communication of the EA is vital. A suitable degree of *formalization* is needed in order to ensure traceable and repeatable results. Furthermore (semi) formalized models and well structured methods are needed to enable division of labor among the stakeholder groups [12, 13]. The general characteristics and purposes of EA are summarized in Table 1.

While documentation and analysis of EA (represented by as-is models) are well covered in academic and practitioner approaches, EA planning is covered much less so far. Since neither the corporation or government agency itself, nor its environment remains static during a transformation project, and because to-be models may change as projects are launched, the consideration of EA dynamics is an important aspect for EA planning.

Table 1. Characteristics and Purposes of EA

Characteristics	Purposes
- Holistic scope - Goal orientation - Formalization	- *Documentation* of organizational structure including artifacts from business and IT and their interrelationships (As-Is architecture), - *Analysis* of dependencies and relationships of As-Is models, - *Planning* and comparing future scenarios (To-Be models), and derivation of transformation projects and programs to achieve a desired EA.

However, as the following section illustrates, the field of EA planning—and information systems (IS) planning in general—is broad and covers very heterogeneous topics. Therefore this contribution focuses on the modeling of EA dynamics in order to support EA planning. In particular, we conceptualize model complexity associated with EA dynamics in business planning and business transformations. In a first step, existing IS/EA planning approaches are analyzed against a holistic, goal oriented and formalized understanding of EA. In a second step, we reflect existing modeling requirements for EA planning from actual industry projects. Based on the findings, a generic EA planning process is proposed, and complexity levels of EA dynamics are described that need to be addressed as EA approaches mature towards a more comprehensive support of EA planning.

In analogy to a process model for design research in information systems [14], this article "identifies a need" for more comprehensive EA planning and lays the foundation for the following "build" phase. However, in this article the respective method artifact is neither built nor is its utility evaluated.

The remainder of this paper is organized as follows. Section 2 presents a literature review of IS/EA planning in the context of to-be modeling and business transformation support. Experience from industry projects in this domain is summarized in section 3. Based on this foundation, requirements for EA planning support including dynamic aspects are derived and subsequently structured by complexity levels in section 4. We then analyze how these complexity levels are addressed by a consolidated EA planning process and which additional steps are necessary. Preliminary results of our research in progress are discussed, and further research activities are proposed in section 5.

2 Literature Review

Only a few contributions to the field of modeling for EA planning support exist so far. However, significant contributions to the broader areas of EA planning and IS planning have been made. Therefore we review not only current literature on EA planning, but also older sources on IS planning which have influenced EA planning.

2.1 IS Planning

Historically, EA planning and to-be modeling evolved from strategic IS planning which was firstly addressed in an MISQ contribution by King in 1978 [15]. This paper proposes a process to design a management information system (MIS) in accordance to the strategy of a corporation or government agency and thereby define a MIS strategy comprising MIS objectives and MIS constraints. As markets, organizational structures and system landscapes added more complexity to the matter of strategic planning and the alignment of business and IT, this approach as well as similar contributions were evolutionarily refined. Strategic enterprise-wide information management [16] and more institutionalized IS planning processes became an issue in the 1990ies [17]. A prominent example for IS planning methods is IBM's Business System Planning (BSP) [18]. BSP aims to (re-)group IT functionalities according to data use and thereby identify application candidates with high internal integration intensity, but limited external interfacing to other applications.

2.2 EA Planning

IS planning and EA planning differ in their approach, goal, and scope. While IS planning is technology driven and refers to the planning of systems (what systems do we need?), EA planning focuses on the business (What do we do now and what do we want to do? What information is needed to conduct our business in the future?) [19]. The offer of new architectural paradigms, such as service orientation, requires for EA planning focusing on supplying information to stakeholders in order to support organizational change.

The term EA planning was first introduced by Spewak, who defines EA planning as "the process of defining architectures for the use of information in support of the business and the plan for implementing those architectures." [19] The underlying understanding of EA covers the whole of data, applications, and technology. Plan—in this context—is referred to as the definition of the blueprint for data, application, and technology as well as the process of implementing the blueprint within an organization. The work of Spewak was updated in 2006, emphasizing the importance of business knowledge and organizational issues during the planning process [20]. Fig. 1 gives an overview of the proposed enterprise architecture planning method that is also referred to as the Wedding Cake Model. A detailed description of the method can be found in [19, 20].

The process steps are to be read top down and from left to right. The definition of the to-be architectures for data, application, and technology are based on their current states as well as on business knowledge that determines major requirements. Yet, the

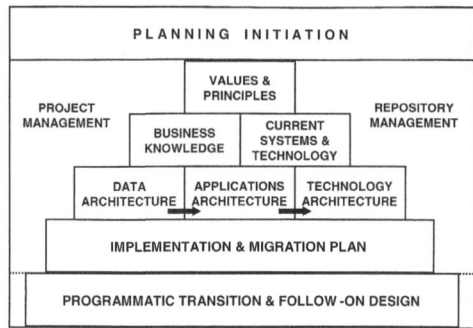

Fig. 1. Wedding Cake Model [19, 20]

process ignores that the business requirements and therefore respective to-be architectures can change as the process is executed.

Based on an extensive literature review, Pulkkinen and Hirvonen propose an EA development process model that guides the incremental stepwise planning and development of EA [21, 22]. Due to the high complexity of the EA development task, only discrete EA development projects are considered. Moreover, the EA process model is intended to cover both EA planning and enterprise systems development.

The authors emphasize the aspect of user participation and decision making in the process which is structured by the EA management grid [23]. Following this two-dimensional grid, the process model proposes that even at the enterprise level technology decisions should be made and then transferred to the underlying levels.

The EA process is further refined in [21] which is depicted in Fig. 2: Pulkkinen proposes parallel domain level decisions implementing the decisions made on the enterprise level (arrow A). After parallel sub-cycles in domain decisions, system level decisions are derived (arrow C). Additionally, the reuse of successful implementations from lower levels to the enterprise level is supported (arrow B). The author

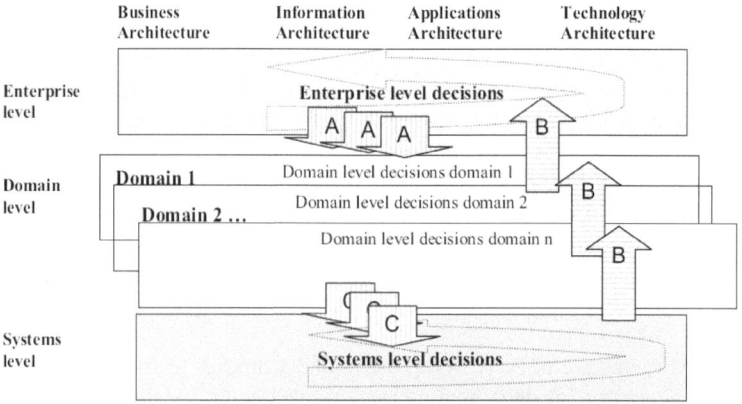

Fig. 2. The Refined EA Process [21]

especially points out the importance of detection of domains within EA development which may result in concurrent planning efforts requiring additional coordination.

Similar approaches for establishing the EA, comprising evaluation, creation and development, are presented by Op't Land et al. [11] and Niemann [24] (cf. Table 2).

Leppänen et al. propose a first step towards an EA planning method by developing a contingency framework that lists several situational factors derived from method engineering and information system development literature [25]. These factors include enterprise/organizational characteristics, persons involved, goals pursued and characteristics of the EA method to be constructed. The contingency model is supposed to support the selection and adaptation of EA method engineering approaches in order to develop a situational method for EA planning.

A specific view on dynamic aspects of EA Planning is presented by Murer et al. under the term "Managed Evolution" [26]. This approach aims at balancing the ratio between the benefits for business and the IT development efficiency. This is realized by using metrics and evaluating both aspects in short time horizons, i.e. for each development project, in order to carefully plan the evolution of large information systems.

2.3 Modeling of EA Planning

The work of Buckl et al. [27] focuses on models for the management of application landscapes with emphasis on temporality aspects, i.e. information about projects that change the application landscape or business changes that affect IS applications. The authors identify three time-related dimensions that need to be considered: an application landscape is planned for a specific time, it has been modeled at a certain time, and different variants of planned landscapes may exist concurrently. Furthermore, five key requirements for supporting temporality aspects in application landscape management are identified. Among these, there are the deduction of future application landscapes from project tasks, the integration of project portfolio management with the application landscape planning process and the possibility to compare variants of future application landscapes. Referring to findings from discussions on object-oriented models, the authors propose the transfer of temporal patterns to model application landscapes considering temporality aspects. As one approach to compare different application landscape models, the evaluation of failure propagation via metrics is presented Lankes et al. in [28] and [29].

The modeling aspect of EA planning is also addressed by EA tool vendors. Based on an extensive survey that analyzes 13 EA management tools, Matthes et al. find that the static complexity of constituents and dependencies is handled well by current vendors, e.g. by providing visualization and collaborative maintenance functionalities [30]. However, dynamic aspects resulting from changes over time are not addressed well by most current EA management tools [31]. While nearly all tools support as-is and to-be modeling, road mapping, versioning and transformation paths are usually not addressed in a sophisticated manner. In addition, EA metrics that would allow for comparison of different to-be scenarios are not covered well in current EA tool support [30]. Also Gartner [32] and Forrester [33] attest a good coverage of niches of dynamic aspects in EA management such as lifecycle management and simulation capabilities. However, there is no EA tool comprehensively addressing dynamic aspects of EA.

2.4 Evaluation

Regarding the premise of a holistic scope of EA, approaches that are restricted to IS or application landscapes cannot be satisfactory (e.g. [19, 20] but also [27, 28, 29] need to be questioned). Findings from IS planning give valuable hints, but have to be significantly extended in order to be useful for EA planning.

Another result of the literature review is that the majority of research results only focuses on EA planning as an unidirectional planning process that aims at improving the current architecture [21, 22, 23, 25]. This includes a defined start date and end date of the process as well as a defined result, i.e. one target architecture for one point in time. In addition, most sources cover individual dynamic aspects such as adaptations of target architecture models to changing conditions, life cycles of individual artifacts, the evaluation of model alternatives or the support of transformation from as-is architecture to to-be architecture. However, there is no comprehensive modeling method for EA planning. Extensions for existing modeling processes for EA planning focusing on dynamic aspects are therefore proposed in section 4.

3 Review of Current Industry Practices

In order to illustrate the need of a more comprehensive approach to modeling for EA planning, we will present two cases from the financial services industry.

3.1 Company A

Company A provides IT outsourcing services and banking solutions. The primary product is an integrated banking platform that is offered to private banks and universal banks. The organization focuses on three main fields, namely application development, application management and operations, and therefore offers an integrated portfolio to its customers. The application development division is responsible for the development of the integrated banking platform. The development activity management is planned and controlled by the architecture team using a home grown solution to create to-be models and manage development projects within the banking platform. This solution combines modeling features and project management capabilities to ensure consistent evolution of the platform. Major challenges within the architectural development plan are the coordination of the activities of the development teams and assurance that milestones of the various integration and development activities are met simultaneously. If, for example, a component of an application needs an interface to a component of another application at a certain time for a certain milestone (e.g. test or release), it has to be assured that both components are available at that very point in time. This simple example grows very complex as the banking platform comprises of over 200 applications, each consisting of a multitude of components that each have their own lifecycles as well as precursor and successor relationships. Generally speaking, the following questions need to be answered within the architectural development plan:

- What are the relationships of architectural elements and what are the impacts of local changes to other elements?
- How can lifecycles of elements and their impacts be modeled?

These dynamic aspects of the architectural development are to some extent supported by a homegrown solution, but yet need strong governance caused by various manual steps in the planning process. This is partially due to specifics of the planning method that is based on implicit knowledge held by individuals.

3.2 Company B

Company B is an internationally operating bank based in Switzerland. During recent decades, mergers led to an increasing complexity of its application landscape. Regarding architecture layers, business architecture (i.e. partly strategy, but mainly organizational artifacts), application and integration architecture, software and component architecture (i.e. software artifacts), and technical architecture (i.e. infrastructure artifacts) are distinguished. Architecture management is realized by more than 90 architects and comprises architecture governance that is enforced in individual IS projects. However, while IT architecture is strong in the bank's home country, the bank has to face challenges due to heterogeneous local solutions in almost every country.

In order to enable a better management of the heterogeneous application landscape, an EA project is currently being conducted. The project focuses on an integrated view on the different solutions the IT departments offer to the company's operating departments and teams worldwide. Such an integrated view should enable solution roadmap planning, too. Therefore, the following questions need to be answered continuously:

- Which projects should be shifted back or forward in order to meet the needs of a certain solution roadmap?
- Which projects affect which lifecycle planning of a certain solution?
- Does postponing of a project affect the lifecycle planning of a certain solution?

An EA approach aiming at these requirements must be capable of consolidating information on different projects affecting solution development, e.g. release planning, component development and customer request management for customized solutions. This approach requires the inclusion of dynamic aspects such as solution and component lifecycles, but especially the support of multi project management. Regarding the actual planning process, the EA approach must support the transformation process from as-is (application) architecture to to-be (application) architecture.

3.3 Implications

Although we provide only two cases from current industry practices here, these examples show the multitude of dynamic aspects that need to be considered during EA planning and evolution. The challenges faced by both companies imply that there is an actual need for an integrated planning method that combines all dynamic aspects and takes into account their interrelationships. For example, company A has identified the need to combine to-be modeling with lifecycles on one hand and the coordination of development activities on the other hand. Similarly, company B is aiming at an alignment of solution roadmap planning and multi project planning. These experiences require the integration of

- project management (organizing programs and individual projects)
- release management (planning development roadmaps), and
- lifecycle management (phases and milestones for different EA elements).

Additionally, complex temporal as well as technical interdependencies between the planning of EA elements, of partial architectures and of projects need to be addressed. The challenge for enterprise architects in the presented cases is to extend the transparency of a current situation provided by as-is EA models to a number of future situations represented by alternative to-be models. Due to the complexity of the interdependent system of as-is as well as alternative to-be models, a sound method supported by EA tools is needed.

Current practices also indicate that not only precise models of one or several target architectures are in use, but that all planning efforts are guided by an "architectural vision". Such a vision serves as a guideline for the architectural evolution, while in most cases there are no ambitions that it will actually be materialized ever. The architectural vision might be deducted from a strategic vision given by business departments or IT departments. It may, for example, specify the substitution of a certain standard software product or platform. The influence of such a vision on the planning process needs to be considered in an integrated concept for EA planning. Ultimately, in order for an EA planning concept to be applicable in practice, it needs to take into account contingency factors like budget, general architecture management guidelines, business support, or legacy architecture.[1]

4   A Concept to Capture Dynamics in EA Planning

The results from the literature review and the review of industry practices (chapters 2 and 3) lead to a set of dynamic aspects which need to be considered and structured along the EA planning process. We therefore propose an EA planning process that is derived and combined from existing approaches. We then propose levels of complexity that structure EA planning dynamics and finally evaluate the process' capabilities to address such levels.

4.1 EA Planning Process

For EA planning purposes, enterprise architects need to know what they are going to plan and how they should proceed in the planning process. Therefore, we firstly derive a generalized EA planning process from respective proposals in literature. Table 2

---

[1] A first step towards the definition of such factors has been done by Leppänen et al. [25] and also Aier et al. [34].

Table 2. Existing EA Planning Processes

Spewak et al. [19, 20]	Niemann [24]	Pulkkinen et al. [21, 22]
A1. Planning initiation	B1. Define goals	C1. Initiation
A2. Define values and principles	B2. Documentation	a. Define goals
	B3. Analysis	b. Resources and constraints
A3. Indentify business knowledge and current systems and technology	B4. Planning alternative scenarios	C2. Planning and development: define needed changes in architectural dimensions
A4. Blueprint data, applications and technology architecture	B5. Evaluation of alternative scenarios	
	B6. Implementation	C3. Ending phase
A5. Develop implementation and migration plan		a. Plan, design and evaluate alterative architectures and solutions
A6. Define programmatic transition		b. Define long term and short term targets

gives an overview of existing approaches. The presented approaches are similar in general yet different in detail. The following process (cf. Fig. 3) condenses the essence of the three approaches described. Subsequently, a more detailed explanation of the steps is given.

Step 1: Define Vision (based on A1, A2, B1, C1a, C1b): Long-term goals and an architectural vision cover the desired state of the architecture that might never be achieved, but delivers the general direction for future plans.

Step 2: Model As-Is Architecture (based on A3, B2): These architectural models serve to document the as-is structure of the organization and are therefore necessary for stakeholder communication and as a planning foundation.

Step 3: Model Alternative To-Be Architectures (based on B3, B4, C3a): Based on the analysis of the as-is architecture, some architecture elements will be more relevant to the planning process than others, e.g. some elements might be subject to higher volatility while others remain stable. Therefore the parts most likely to be changed must be identified, and to-be models depicting the desired changes can be created. Since the architectural vision might be approached in multiple ways, multiple to-be architectures will be built during this phase. Some of these to-be architectures are alternative to each other while some are related to different points in time.

Step 4: Analyze and Evaluate Alternative To-Be Architectures (based on B3, B5, C3a): In order to plan the next state of the current as-is architecture, one of the alternative to-be architectures needs to be chosen. This selection process requires an analysis, evaluation and comparison of the given options.

Step 5: Plan Transformation from As-Is to To-Be Architecture (based on A5, C3a): After the desired to-be architecture is identified, the detailed planning for the transformation process can take place. This step involves project and program planning as current or planned development projects may affect mutual architectural elements. Furthermore, project interrelationships should be identified in order to consolidate projects and resources.

Step 6: Implement Transformation (based on A6, B6): Lastly, the transformation has actually to be implemented. When this step is completed, one of the to-be models becomes the as-is model and the next iteration of the process starts. As this paper focuses on modeling aspects rather than on implementation, this step is not regarded in the following sections.

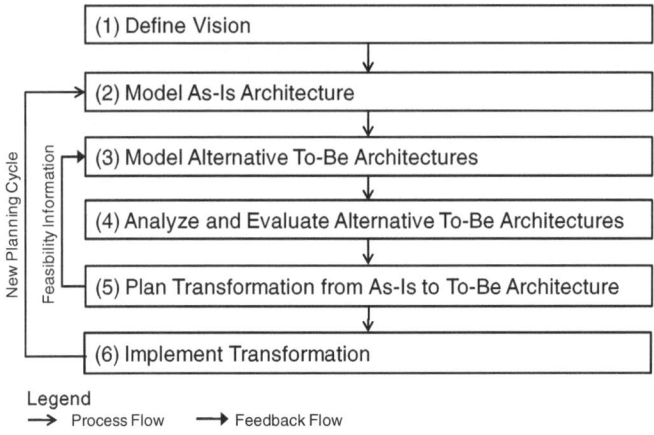

Fig. 3. EA Planning Process

The continuity of time causes dynamic changes within the EA planning process. This especially influences the actuality of the to-be architecture models: Conditions for certain decisions might change with proceeding time. Therefore the point in time when the model was created should be captured as well [27].

During the detailed planning for the transformation process, further knowledge about the possible future states of the as-is architecture may arise. This information should be re-integrated in the modeling process (step 3). The same can occur while conducting the project and in program management: if, for example, concurrent use of resources is detected, this information needs to be fed back into step 3, too. Both aspects have continuous influence on the "decision tree" that is generated by modeling different to-be architectures.

4.2 Complexity Levels in EA Planning

The general temporal influences on the EA planning process result in high complexity that appears while putting the planning process into action. In order to capture this complexity and address open issues from the case studies in section 3, we break down the complexity into different dynamic aspects. On this basis, we distinguish different levels of complexity in EA planning (cf. Table 3).

The first level comprises an *as-is model*, a *to-be model* and, according to step 1 from the EA planning process presented in chapter 4.1, an *architectural vision*. On level 2, the *transformation plan* connecting the as-is state and the to-be model is added. Levels 3 and 4 incrementally include the modeling of *alternative to-be models* and their *comparability*. *Multi-step to-be modeling* and *transformation planning* is regarded from level 5 on. The continuous influence of time and consequential changes like *unplanned amendments* of to-be models are included in level 6 and level 7, while the latter additionally considers further effects on multi-step planning.

Table 3. Levels of Dynamic Complexity in EA Planning

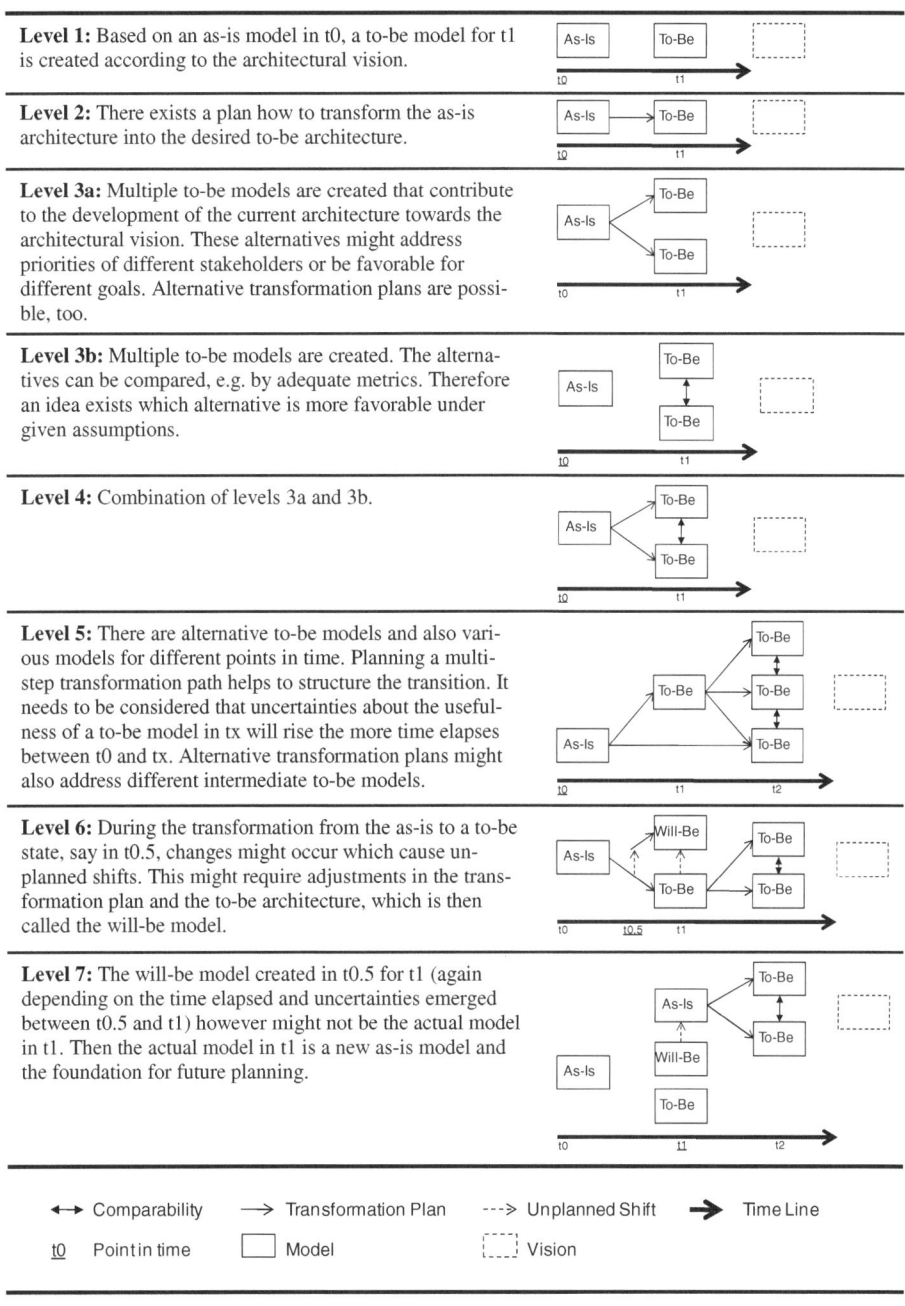

Level 1: Based on an as-is model in t0, a to-be model for t1 is created according to the architectural vision.

Level 2: There exists a plan how to transform the as-is architecture into the desired to-be architecture.

Level 3a: Multiple to-be models are created that contribute to the development of the current architecture towards the architectural vision. These alternatives might address priorities of different stakeholders or be favorable for different goals. Alternative transformation plans are possible, too.

Level 3b: Multiple to-be models are created. The alternatives can be compared, e.g. by adequate metrics. Therefore an idea exists which alternative is more favorable under given assumptions.

Level 4: Combination of levels 3a and 3b.

Level 5: There are alternative to-be models and also various models for different points in time. Planning a multi-step transformation path helps to structure the transition. It needs to be considered that uncertainties about the usefulness of a to-be model in tx will rise the more time elapses between t0 and tx. Alternative transformation plans might also address different intermediate to-be models.

Level 6: During the transformation from the as-is to a to-be state, say in t0.5, changes might occur which cause unplanned shifts. This might require adjustments in the transformation plan and the to-be architecture, which is then called the will-be model.

Level 7: The will-be model created in t0.5 for t1 (again depending on the time elapsed and uncertainties emerged between t0.5 and t1) however might not be the actual model in t1. Then the actual model in t1 is a new as-is model and the foundation for future planning.

4.3 Evaluating the EA Planning Process for Complexity Levels

This section investigates how the proposed EA planning process supports dynamic complexity. Therefore it is analyzed which process steps address the dynamic aspects associated with the different levels of dynamic complexity.

Table 4. Dynamic Aspects in the EA Planning Process

	Step 1: Vision	Step 2: As-Is Architecture	Step 3: Multiple To-Be Models	Step 4: Evaluation of To-Be Models	Step 5: Transformation Plan
Vision	●	●	○	○	○
As-Is Architecture	○	●	○	○	○
To-Be Architecture	○	○	●	○	○
Transformation Plan	○	○	●	○	●
Alternative To-Be Architectures	○	○	○	●	○
Comparability	○	○	○	●	○
Multi-Step Transformation	○	○	○	○	○
Unplanned Shifts	○	○	○	○	○

Legend   ● step explicitly addresses dynamic aspect   ○ step does not address dynamic aspect

This evaluation shows that most dynamic aspects can currently be supported by the EA planning process. But multi-step transformation, the consideration of unplanned shifts and the deviation of models from reality are not addressed yet. If applied to the levels of complexity, this means that levels 1–4 can currently be supported by the steps included in the EA planning process. Levels 5–7 however require an extension of the proposed EA planning process. The following discussion describes the requirements and Fig. 4 depicts a proposal for an extension of the EA planning process.

Level 5 describes a multi-step "decision tree" that demands for the selection of one transformation path. In order to support planning over longer periods of time, the planning process also needs to address intermediate to-be models as partial results as well as the comparison and selection of different combinations of multiple subsequent to-be models. This can reflected in the new process steps 3b and 4b in Fig. 4.

For a realization of levels 6 and 7 the EA planning process must also address the adjustments of to-be models in will-be models due to unplanned shifts during the transformation. Furthermore, respective changes in transformation plans that have an effect on future to-be models need to be covered. Such unplanned changes might also originate from aspects that cannot be modeled but influence any kind of planning process: politics or budgets, for example. Applied to the EA planning process, unplanned changes affect the process steps 2 and 5 (cf. Fig. 4). Finally, these influences trigger feedback to the modeling and comparison of to-be models which need to be adjusted (cf. "Feedback" arrows).

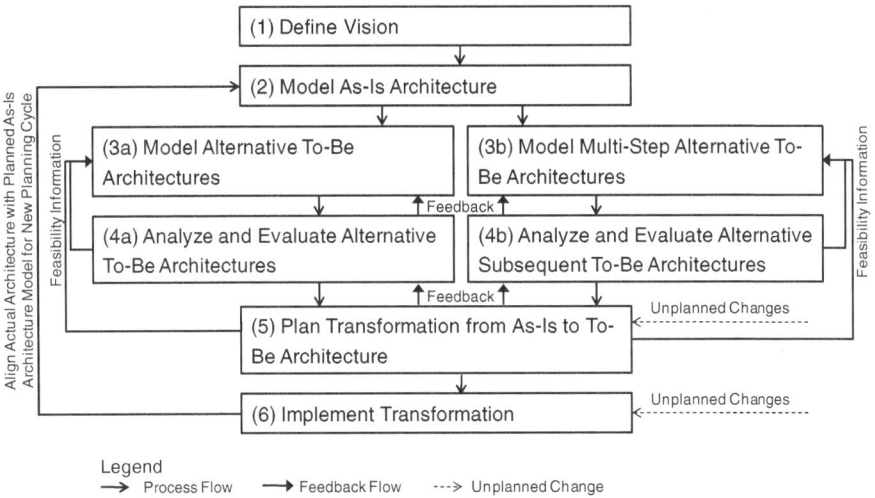

Fig. 4. Extended EA Planning Process Proposal

5   Discussion and Conclusion

This paper presents research in progress related to capturing dynamic aspects in EA planning. To that means, existing approaches in literature as well as open questions from practitioner projects have been analyzed. After consolidating an EA planning process on this basis, we have analyzed how different complexity levels of dynamics are addressed by the different steps of the planning process. This analysis implies extensions to the proposed process model. Multi-step planning as well as capturing differences between models and unplanned changes that have an impact on the planning results are not yet fully addressed. Future research activities will investigate how these aspects can be incorporated into the planning process. Open questions to be answered in this context include:

- Is there a desired level for dynamic complexity in EA planning?
- How can the tradeoffs between (a) pragmatic solutions that only partly consider the complexity levels and (b) sophisticated solutions that include all complexity levels but potentially cause higher planning efforts be addressed?

Furthermore the individual steps of EA planning process need to be detailed. Therefore existing methods for the different tasks will be reviewed in order to analyze their capabilities to capture dynamic aspects in EA planning. This may finally lead to a comprehensive method for EA planning addressing dynamic aspects. Open questions to be answered in this context include:

- How do we decide what architectural elements are relevant for planning, and are therefore part of to-be models?
- How do we capture dynamics of EA models and artifacts by graphical representations?
- What are relevant dimensions and methods for the evaluation of to-be models?

- What lessons can be learned from project and program management to structure the transformation process?
- What are the requirements for an integrated planning method?

In regard to the proposed complexity levels, a comprehensive method for EA planning should be adaptive: Depending on the desired level of complexity to be addressed, the method should provide a situational solution. In analogy to the process model for design research, the method artifact should then also be evaluated, for example, by industry cases.

References

1. IEEE: IEEE Recommended Practice for Architectural Description of Software Intensive Systems (IEEE Std 1471-2000). IEEE Computer Society, New York, NY (2000)
2. Tyler, D.F., Cathcart, T.P.: A structured Method for Developing Agile Enterprise Architectures. In: Proceedings of International Conference on Agile Manufacturing (ICAM 2006), Norfolk, Virginia, USA, pp. 1–8 (2006)
3. Rohloff, M.: Framework and Reference for Architecture Design. In: Proceedings of 14th Americas Conference on Information Systems (AMCIS 2008), Toronto (2008)
4. Winter, R., Fischer, R.: Essential Layers, Artifacts, and Dependencies of Enterprise Architecture. Journal of Enterprise Architecture 3(2), 7–18 (2007)
5. Jonkers, H., Lankhorst, M.M., Doest, H.W.L., Arbab, F., Bosma, H., Wieringa, R.J.: Enterprise architecture: Management tool and blueprint for the organisation. Information Systems Frontiers 8, 63–66 (2006)
6. Lankhorst, M.: Enterprise Architecture at Work: Modelling, Communication and Analysis. Springer, Berlin (2005)
7. Dietz, J.L.G.: Enterprise Ontology – Theory and Methodology. Springer, Heidelberg (2006)
8. Winter, R., Bucher, T., Fischer, R., Kurpjuweit, S.: Analysis and Application Scenarios of Enterprise Architecture – An Exploratory Study. Journal of Enterprise Architecture 3(3), 33–43 (2007)
9. Niemi, E.: Enterprise Architecture Stakeholders – A holistic view. In: Proceedings of 13th Americas Conference on Information Systems (AMCIS 2007), Keystone, CO (2007)
10. Ylimäki, T.: Potential Critical Success Factors for Enterprise Architecture. Journal of Enterprise Architecture 2(4), 29–40 (2006)
11. Op't Land, M., Proper, E., Waage, M., Cloo, J., Steghuis, C.: Enterprise Architecture – Creating Value by Informed Governance. Springer, Berlin (2009)
12. Frank, U.: Perspective Enterprise Modeling (MEMO) – Conceptual Framework and Modeling Languages. In: Proceedings of 35th Hawaii International Conference on System Sciences (2002)
13. Jonkers, H., Lankhorst, M., van Buuren, R., Hoppenbrouwers, S., Bonsangue, M., van der Torre, L.: Concepts for Modelling Enterprise Architectures. International Journal of Cooperative Information Systems 13(3), 257–287 (2004)
14. Rossi, M., Sein, M.K.: Design Research Workshop: A Proactive Research Approach (2003), http://tiesrv.hkkk.fi/iris26/presentation/ workshop_designRes.pdf (last access 01.02.2004)
15. King, W.: Strategic Planning for Management Information Systems. MIS Quarterly 2(1), 27–37 (1978)
16. Targowski, A.: The architecture and planning of enterprise-wide information management systems. Idea Group Publishing, Harrisburg (1990)

17. Eliot, L.B.: Information systems strategic planning. Computer Technology Research Corporation, Charleston (1991)
18. IBM: Business Systems Planning: Information Systems Planning Guide, 4 edn., vol. 1, Atlanta, IBM (1984)
19. Spewak, S.H., Hill, S.C.: Enterprise Architecture Planning – Developing a Blueprint for Data, Applications and Technology. John Wiley & Sons, New York (1993)
20. Spewak, S.H., Tiemann, M.: Updating the Enterprise Architecture Planning Model. Journal of Enterprise Architecture 2(2), 11–19 (2006)
21. Pulkkinen, M.: Systemic Management of Architectural Decisions in Enterprise Architecture Planning. Four Dimensions and Three Abstraction Levels. In: Proceedings of 39th Annual Hawaii International Conference on System Sciences (HICSS 2006), Honolulu, Hawaii, pp. 179a (1–9). IEEE Computer Society, Los Alamitos (2006)
22. Pulkkinen, M., Hirvonen, A.: EA Planning, Development and Management Process for Agile Enterprise Development. In: Proceedings of 38th Annual Hawaii International Conference on Systems Sciences (HICSS 2005), pp. 223.3 (1–10). IEEE Computer Society, Los Alamitos (2005)
23. Hirvonen, A., Pulkkinen, M.: A Practical Approach to EA Planning and Development: the EA Management Grid. In: Proceedings of 7th International Conference on Business Information Systems, Wydawnictwo Akademii Ekonomicznej w Poznaniu, pp. 284–302 (2004)
24. Niemann, K.D.: From Enterprise Architecture to IT Governance. Elements of Effective IT Management. Vieweg, Wiesbaden (2006)
25. Leppänen, M., Valtonen, K., Pulkkinen, M.: Towards a Contingency Framework for Engineering an Enterprise Architecture Planning Method. In: Proceedings of 30th Information Systems Research Seminar in Scandinavia (IRIS 2007), Tampere, Finland (2007)
26. Murer, S., Worms, C., Furrer, F.: Managed Evolution – Nachhaltige Weiterentwicklung grosser Systeme. Informatik-Spektrum 31(6), 537–547 (2008)
27. Buckl, S., Ernst, A., Matthes, F., Schweda, C.M.: An Information Model for Landscape Management – Discussing Temporality Aspects. In: Proceedings of 3rd Workshop on Trends in Enterprise Architecture Research (TEAR 2008), Sydney (2008)
28. Lankes, J., Schweda, C.: Using Metrics to Evaluate Failure Propagation in Application Landscapes. In: Proceedings of Multikonferenz Wirtschaftsinformatik (MKWI 2008), München (2008)
29. Lankes, J., Matthes, F., Ploom, T.: Evaluating Failure Propagation in the Application Landscape of a Large Bank. In: Proceedings of Component-Based Software Engineering and Software Architecture (CompArch 2008), Karlsruhe (2008)
30. Matthes, F., Buckl, S., Leitel, J., Schweda, C.M.: Enterprise Architecture Management Tool Survey 2008, München: Software Engineering for Business Information Systems (sebis) Ernst Denert-Stiftungslehrstuhl Chair for Informatics 19 TU München (2008)
31. Buckl, S., Dierl, T., Matthes, F., Ramacher, R., Schweda, C.M.: Current and Future Tool Support for EA Management. In: Proceedings of Workshop MDD, SOA und IT-Management (MSI 2008), Oldenburg, Gito (2008)
32. James, G.A., Handler, R.A.: Cool Vendors in Enterprise Architecture, Gartner Research G00146330 (2007)
33. Peyret, H.: Forrester Wave: Business Process Analysis, EA Tools, And IT Planning, Q1 2009, Forrester Research (2009)
34. Aier, S., Riege, C., Winter, R.: Classification of Enterprise Architecture Scenarios – An Exploratory Analysis. Enterprise Modelling and Information Systems Architectures 3(1), 14–23 (2008)

An Approach for Creating and Managing Enterprise Blueprints: A Case for IT Blueprints

Pedro Sousa[1,2,3], José Lima[1], André Sampaio[1], and Carla Pereira[2,3]

[1] Link Consulting, SA
[2] Instituto Superior Técnico (IST), Technical University of Lisbon
[3] Inov-Inesc, Lisbon, Portugal
{pedro.sousa,jose.lima,andre.sampaio}@link.pt,
carla.pereira@inov.pt

Abstract. One important role of Enterprise Architecture aims at modeling enterprise artifacts and their relationships, ranging from the high-level concepts to physical ones such as communication networks and enterprise premises. As it is well known, these artifacts evolve over time, as well as their relationships. The dynamic nature of such artifacts has been a difficulty not only in modeling but also in keeping enterprise blueprints updated. This paper presents our approach to handle blueprints of the Enterprise Architecture, based on several years and projects in large organizations, both in the financial and telecommunication industry.

We started by considering "projects" as the changing elements of Enterprise artifacts and achieve a scenario where blueprints are automatically generated and updated, and a time bar allows traveling from the past (AS-WAS), to the present (AS-IS) and to the future scenarios (TO-BE). The paper also presents an overview of the underlying model, the applied methodology and the blueprints that we found to be a valuable instrument amongst elements of different communities: Project Management, IT Governance and IT Architecture. In spite that the cases studies are targeted to the IT domain, the lessons are valid for other architectural areas.

Keywords: IT Blueprints, Enterprise Architecture, IT Governance.

1 Introduction

As in any complex system, enterprises would be better understood if one could have a blueprint (schematic representation) of the artifacts that constitute the organization and their relations. In the IT domain, blueprints have always been perceived as an important asset, especially by the IT Architecture teams or departments. In fact, many companies have been trying to make blueprints of the IT landscape, from high level maps to detailed ones. But the truth is that companies fail to have such maps, claiming that update costs are simply too high given the rate of changes of the organization artifacts.

Blueprints come in many shapes and detail levels. In order to clarify what we mean by a "blueprint", we present two examples of different level of detail and scope. On

A. Albani, J. Barjis, and J.L.G. Dietz (Eds.): CIAO!/EOMAS 2009, LNBIP 34, pp. 70–84, 2009.
© Springer-Verlag Berlin Heidelberg 2009

the left side of figure 1, we present one example of a very high level business view of retail banking, following the classification schema proposes in [1]. On the right side we present a typical application landscape with interactions between applications.

We have been doing professional consultancy in the domain of Enterprise Architecture for over a decade and have found many other reasons, other than costs, that are preventing companies to have blueprints up-to-date. Probably the most common and simple reason is that, quite often, IT professionals assume that the adoption of a given modeling notation (i.e. UML) is enough to start producing blueprints. But it is well known that, in fact, behind each blueprint there is a *theory* that defines the governing rules, a *model* that identifies the properties and semantics of artifacts, and a *notation* to graphically express such artifacts, and also a *problem*, to provide a purpose of each blueprint and thus, making it possible to decide what artifacts should appear in each one [2].

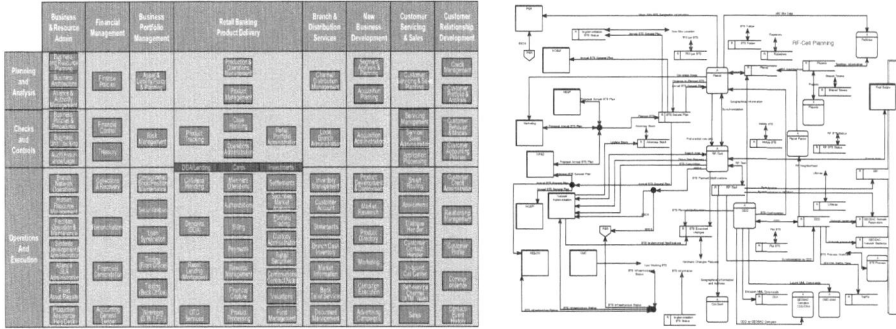

Fig. 1. Blueprint Examples

This paper is not about the right or the best *theory*, *model* or *notation* for enterprise architecture. It is about what needs to be added to known theories, models and notations to allow blueprints to be systematically produced and maintained in real organizations.

This paper is structured as follows: in the next section we present a more detailed view of the problem, trying to narrow down the key issues behind the difficulty of creating and maintaining blueprints; in section 3 we present the related work in the scope of Enterprise Architecture; in section 4 we present some aspects of the theory and model necessary to support our approach; in section 5 we present the methodology used in real cases; in section 6 we present the BMS, our software solution for of blueprint management, and finally, we conclude in section 7.

2   Problem Clarification

In order to keep blueprints up-to-date, one needs two basic things:

- Information about what has changed.
- Rules to update the blueprints accordingly.

Regarding the first issue, we have found that such information normally exists in the notebooks and agendas of the IT professionals, namely of those that were involved in the changes. However, these notes were written with the purpose to refresh the memory of those who have written them. We could envisage that the problem would be solved if IT professionals use standard concepts and notations in their personal notes and published them into a common repository at the end of the day. But in medium and large size organizations, IT reality is changed mostly by IT projects, and therefore IT projects are the best entity to report back the changes in some normalized form.

But managed IT projects do have a work plan to produce the intended artifacts. For example, if a project intends to make system A that sends data to an existing system B, then both the system A and the data flow to B should be referenced in the project plan details.

Therefore, the real questions come down to:

Q1. Are the IT artifacts referred in IT project planning the ones that appear in IT architectural blueprints?

Q2. Are project plans up-to-date enough so they can be a trustful source of information?

A second concern, still related with the first issue, is the extra complexity that enterprise wide architecture blueprints may bring to IT projects. If fact, since enterprise wide blueprints tend to focus on global views of IT artifacts, rather than on the subset of artifacts that are relevant for a given project, reading and updating enterprise blueprints is more complex than it could be. For example, if a project only has to be concerned with 100 artifacts out of a total of 1000 artifacts, then the project should handle a 100 artifact's blueprint rather than a 1000 artifact's blueprint. Given that IT projects are mostly stressed for time and budget, such additional complexity is also a critical aspect, in particular if they have to keep blueprints updated.

Therefore, other relevant questions are:

Q3. Can we provide to each IT project a blueprint no more complex than it should be?

Q4. Can changes in such blueprints be automatically propagated back to enterprise-wide and more complex blueprints?

Regarding the second issue, the fact is that today´s blueprints are mostly a hand-made piece of art, made with a mouse; likewise a painter uses a brush. A blueprint is mostly a personal achievement, not an industrialized result. Most concepts depicted in IT blueprints such as Information Systems, Applications, Platforms, Application Components, Nodes, amongst many others, are not at all clear amongst IT professionals, as one could expect. Unlike with professionals from other engineering domains, when faced with a given reality, different IT professionals name and classify the same artifacts differently. The use of a common and well known notation such as UML does not help at all on the fundamental aspects, because the semantic meaning of symbols is not defined, and must be a personal decision. Furthermore, there is no clear definition to what artifacts should be represented in each blueprint. Once again, it is up to the designer to decide what is relevant and what is not relevant to represent/model in each blueprint.

Therefore, regarding the second issue, the relevant questions are:

Q5. To what level of detail/semantic should one define artifacts and concepts?
Q6. How can one decide which architectural blueprints to use and what artifacts should each blueprint represent?

3 Related Work

Blueprints and schematic representation are common ways of communication between people, namely to express an architectural description of things, like a system, an object, a model or, in our case, an Enterprise.

As clearly described in the IEEE1741 [2], as well as in other works [1, 3, 4], behind a architectural description, there is always a set of other concepts that must be defined to ensure that the architectural description is properly understood, namely a notation, a concern, a model, a view and viewpoint. A considerable amount of effort has been put in the development of these concepts, and in fact most EA frameworks do propose a set of concerns and corresponding views and viewpoints, a model, concepts and in some cases even a notation [2, 3, 5-9]. It also commonly assumed that, architects are able to produce blueprints and models based on proposed EA frameworks, which in turn, sustain developments in many other areas, such as strategic alignment [10], IT Governance [11], Application and Project Portfolio Management [12-14] to name a few. In [15], one my find an overview of uses and purposes of EA.

But the assumption that an architect (or an army of them) is able to model the organization is a valid one only for the domains where the change rate is in fact low, as for example the Enterprise Ontology [16], the enterprise organic structure or the enterprise vision and mission. For the domains with a high rate of changes, such as business processes or IT, one cannot assume to have valid blueprints, simple because the effort to keep them up-to-date is too high. In other words, the current EA frameworks and models do not capture the dynamic nature of enterprises.

To our knowledge, the problem of creating and keeping blueprints up-to-date in an automatic manner has not been an issue in the EA community. As referred in chapter 2, two main issues need to be addressed.

The first - information about what has changed – concerns mostly with establishing an information flow, from the ones making the changes in the IT, to the ones updating the blueprints. We found related work in the area of IT portfolio and project management [12-14], where EA is used as a source of information to feed decisions in IT portfolio and project management, but they do not established a detailed update from IT projects to EA.

The second - the rules to update the blueprints accordingly – concerns mostly on how to overcome is the lack of semantic behind common notations (as UML, SysML, IDEF, amongst many others). As Mark Lankorst refers, they are mostly symbolic models, not semantic ones [3]. Archimate [3] moves one step forward by providing a stronger model, but true semantics requires a theory, as the Ψ-theory [17], to sustain models and methodologies and well understood blueprints. In what concerns the IT, we have knowledge of a semantically sound model, although some progress has been made [18].

4 Fundamentals of Our Approach

The following description intends to give the reader the basic grounds for the practical work presented in this paper. It does not intend to be a full explanation of the model, and many aspects of it are not referred.

Let an Enterprise E be modeled as a graph G of artifacts and their relationships, the vertices and the edges of the graph accordingly. Let A and R be the set of all artifacts and relationships[1], accordingly. Since this graph changes over time, let G_t (A, R) represent the value of G at time t, where t is a discrete variable corresponding to the succession of states $\{G_0, ..., G_n\}$ of G_t (A, R).

Let each artifact $a \in A$ have a type $y \in \Gamma$, where Γ is the set of all types. The statements "a IsA y" and "$y= type\ (a)$" state that y is the type of artifact a. A type defines the properties and possible values for artifacts of that type. Relationships are typed after the types of connected vertices[2][19]. For clarity sake, we'll represent types in italic and with the first letter in capital (e.g. $Type$) and instances in italic.

We start by introducing two fundamental types of Γ:

- *Blueprint,* whose instances contain references to others artifacts. A given *artifact* is represented on a given *blueprint* if graph G holds as a relation between them.
- *Project,* whose instances contain references to artifacts related with the project.

We further define the state_of_existence of all artifacts other than *Blueprint* as one of the following states:

- **Conceived:** If it is only related with *blueprints.*
- **Gestation:** If it is related with alive *projects* and is not related with any other artifacts other than *blueprints.*
- **Alive:** If it is related with other artifacts in the <u>alive</u> state. This means that it may act upon other artifacts in <u>conceived</u>, <u>gestation</u> or <u>alive</u> states.
- **Dead:** If it is no longer in the <u>alive</u> state[3].

Let *Project.aliveList* and *Project.deadList* be the list of artifacts to become <u>alive</u> and <u>dead</u> during the project.

We now come back to the set of states $\{G_0, ..., G_n\}$ of G_t (A, R) that represent the sequence of states of the organization, and consider the sequence of <u>alive</u> state of the organization $\{G_{A0}, ..., G_{An}\}$. Whenever a *project* ends, the set of <u>alive</u> artifacts in the enterprise changes from a state G_{An} to the a state G_{An+1} where

$$G_{A_{n+1}} = G_{A_n} \cup project.aliveList \setminus project.deadList$$

---

[1] Notice that between two given artifacts there may exist relationships of different types.

[2] The simple fact that relationships have a type means that: (i) the set Γ includes also the types of all relationships, and (ii) the relationships are in fact "processors" in the context of General System Theory [19], as are the artifacts.

[3] In the context of General System Theory [19], artifacts in the dead state are necessary passive objects/systems, regardless of the level they had while alive (from 1-passive- to 9 - finalizing).

This allows us to move back and forth in time from the past to the present and from the present to the future. In particular, it allow us to define the *ToBe(t)* as the set of alive artifacts at time *t* based on the *AsIs* state, namely:

$AsIs = G_{A_s}(A, R);$

$ToBe(t) = \ G_{A_s}(A, R) \cup \{ p. aliveList \mid \forall p{:}\, p \; IsA \; Project \; \wedge \; s \le p. endTime \le t \; \} \; \setminus$
    $\{ q. deadList \mid \forall q{:}\, q \; IsA \; Project \; \wedge \; s \le q. endTime \le t \};$

Where, *s* is the time of *AsIs* state (presumably the current date) to and *p.endTime* is planned time for the project *p* to end. It is also clear that the number of possible states of G_A between any given points in time corresponds to the number of projects that end between those two points.

We now focus on a particular type of artifacts: the artifacts of type *System*. We start by clarifying the relations[4] "IS_PART_OF" and "ACTS_UPON", according to notation used in [16]:

- IS_PART_OF represented as " $\prec$", is a relation between two artifacts such that: for any two artifacts (x, y), $x \prec y$ if and only if, in order for *y* to be <u>alive</u> *x* must also be <u>alive</u> and *x* cannot be part of another artifact that *y* is not also part of.
- ACTS_UPON represented as "$\rightarrow$", is a relation between two artifacts such that: for any two artifacts(x, y), $x \rightarrow y$ if and only if, *x* causes changes in state/behavior of artifact *y*. This implies that *y* is not in the <u>dead</u> state.

Following the description presented in [16], a system σ is defined by its *Composition*, *Environment* and *Structure*:

- The *Composition C* of a system σ is the set of artifacts that:

$$C(\sigma) = \{x{:}\, x \; IsA \; Component \wedge \; x \; \prec \sigma\}$$

- The Environment E of a system σ is the set of artifacts that:

$$E(\sigma) = \{x{:}\, x \; \notin \; C(\sigma) \wedge \exists y{:}\, y \in C(\sigma) \; \wedge (x \rightarrow y \vee y \rightarrow x)\}$$

- The Structure S of a system σ is the set of related artifacts defined as:

$$S(\sigma) = \Big\{ <x, y> \mid (x \rightarrow y \vee y \rightarrow x)$$
$$\wedge \Big(x, y \in C(\sigma) \vee \big(x \in C(\sigma) \wedge y \in E(\sigma)\big)\Big)\Big\}$$

Notice that, in spite that the above expressions are time invariant, the set of artifacts that belong to $C(\sigma), E(\sigma)$ and $S(\sigma)$ do change over time, since the IS_PART_OF and ACTS_UPON relations are defined over the <u>alive</u> state of artifacts, which change with the projects alive and dead lists.

After the definition of artifacts of type *System*, we are able to define its basic views (Organic, Environment, Composition and Structure), used by our blueprint engine to produce the blueprints:

---

[4] Both relations are transitive, but we will not explore such properties in this paper. We consider only the direct dependencies.

- The Organic View V_{Org} depicts the artifacts of subtype of *System* in a hierarchically manner according to the value of a given property P_{org}.

$$V_{ORG}(T_{Org}, OrgHierarchySet, t)$$
$$= \{Depiction(\sigma) \mid \forall \sigma: type(\sigma)$$
$$\in T_{Org} \land P_{org}(\sigma) \in_t OrgHierarchySet\}$$

Where is the set of subtypes that will be depicted, *OrgHierarchySet* is the hierarchy values according to which artifacts will be graphically arranged, and $P_{org}(\sigma) \in_t OrgHierarchySet$ states that the value of P_{org} of property σ that must be one of possible state defined in *OrgHierarchySet* at time t.

- The Environment View V_{ENV} of a system depicts the artifacts that belong to the Environment of that system.

$$V_{ENV}(\sigma, T_{ENV}, t) = \{Depiction(x) \mid \forall x: type(x) \in T_{ENV} \land x \in_t E(\sigma)\}$$

Where T_{ENV} is the set of types that will be depicted, and $x \in_t E(\sigma)$ states that artifacts must belong to environment of σ at time t.

- The Composition View of a system depicts the artifacts that belong to the system *Composition* and their relationships.

$$V_{COMP}(\sigma, T_{STR}, t) = \{Depiction(x) \mid \forall x: type(x) \in T_{COMP} \land x \in_t C(\sigma)\}$$

Where T_{COMP} is the set of types that will be depicted, and $x \in_t C(\sigma)$ states that artifacts must belong to Composition of σ at time t

- The Structure View of a system depicts all the relationships between any two artifacts related with the system.

$$V_{STR}(\sigma, T_{STR}, t) = \{Depiction(x, y) \mid \forall x, y: type(x), type(y)$$
$$\in T_{STR} \land (x, y \in_t S(\sigma) \lor (x \in_t S(\sigma) \land y \in_t E(\sigma)))\}$$

Where T_{STR} is the set of types that will be depicted, and $x, y \in_{ti} S(\sigma)$ states that artifacts x, y belong to Structure of σ at time t.

4.1 The Case for IT Artifacts

The application of the above model implies the definition of type hierarchy Γ, making clear the relevant concepts and their relationships. As in most Enterprise Architecture frameworks, IT related types include concepts, such as: *BusinessProcess*, *InformationEntity*, *Stakeholder*, *Repository*, *DataFlow*, *Service*, *Domain*, *Solution*, *Application*, to name a few. In most cases, these concepts are defined too loosely, to ensure a full and unique understanding amongst different persons, especially if they come from different communities. Therefore, the further we close artifact concepts the easier the communication gets. We present the example of applications and their components.

Let application to be a *System*, whose *Composition* is a set of artifacts of type *AppComponent*, being the later a subtype of *Component*, which is in turn a subtype of *System*[5].

---

[5] Therefore *AppComponent* is also a *System*. The recursive aspects will not explored in this paper. Suffice is to say that *Domain*, *Solution*, *Application* and *AppComponent* are all subtypes of *System* and are implemented as instances of the same class.

According to definition of *System*, many sub-systems can be considered and defined for a particular system. Therefore, the identification of *AppComponent* of a given application is normally a personal decision, with unclear rules and assumptions. Such degree of freedom makes blueprints unclear because one does not know the reasons why a particular set of *AppComponents* were considered, instead of another set. Thus, in order to produce blueprints more clear, we were forced to establish additional rules guiding the finding of the proper set of *AppComponents*.

We present a definition that has proven to be useful and simple. Let *Application* to have a layered structure[6]. For the sake of simplicity, let us assume the traditional layers: *UserInterface*, *BusinessLogic*, *Integration* and *Data*. Let *ITPlatform* artifact to be also of type a System.

A given artifact *c* is a *AppComponent* of given *Application a* is and only if it complies with the following conditions:

> (i) *c* is a subsystem of *a*. This means that[16]:

$$C(c) \subseteq C(a)$$

$$E(c) \subseteq C(a) \setminus C(c) \cup E(a)$$

$$S(c) \subseteq S(a)$$

> (ii) *c* is related with one and only one layer of *a*. Let *p,q* be two application layers of *a* . This means that:

$$\forall x: \ x \in E(p), \ \nexists q \neq p: y \in E(q) \ \wedge \ y \in C(c)$$

> (iii) *c* is related with one and only one *ITPlatform*, the platform where the *AppComponents* perform/execute. The formulation is the same as the previous, considering *p,q* to be *ITPlatforms*.

Altogether, these conditions state that an application component is a system executing on a given platform to perform one application role (*UserInterface*, *BusinessLogic*, *Integration* and *Data*). If a more fine grained rule is required, one normally add fourth rule regarding business functions, increasing further the definition of a component: an application component is a system executing on a given platform to perform one application role of a business function.

We now consider some simplifications based on two key aspects that we found to be true in some large companies, in which we found that:

- IT project management is a mature discipline, meaning that *Project* artifacts are well established and managed ones.
- IT production environment is a managed asset, meaning that placing IT artifacts into production is also a mature discipline.

Under such conditions, it is a reasonable assumption that an IT artifact:

- Is in the gestation state when it is being developed within a given IT project.
- It becomes in the alive state, when they are placed onto the production environment as a result of some IT project.

---

[6] A similar construction could be done for a service oriented structure.

- It becomes in the <u>dead</u> state, when they are removed from production environment as a result of some IT project.

Regarding IT projects, we also consider that:

- Artifacts in project <u>alive</u> and <u>dead</u> lists become <u>alive</u> or <u>dead</u> at the end of the project.
- Project <u>alive</u> and <u>dead</u> lists are up-to-date at least in two moments in time: when the project starts and when the project ends. In the first moment, these lists are a promise of what the project intends to do, and in the second moment theses lists are what the project actually did.

This means that project <u>alive</u> and <u>dead</u> lists will be considered as *ToBe* until the project has ended, and will become *AsIs* at project termination, accordingly to the expressions stated before.

5 Methodology

In order to apply the previous model to organizations we follow the following phases:

1. **Problem Statement.** We identify the key concerns of each participating communities, namely IT projects, IT Architecture and IT Governance, so that blueprints can be designed in a way such that they are useful for the above communities, and concepts may be defined in the appropriate level of detail. Thus, in this first phase, we establish desired goals and outputs of the project.
2. **Information and Processes analysis.** One analyzes the actual processes of Project Management, IT architecture and IT Governance, in the organization and the information exchanged between them. This allows us either to confirm the expectations rose in the previous phase or to adjust to them accordingly.
3. **Artifact definition.** One revises the definition of the artifacts required to address the concerns identified in the previous step, and clarifies the source information and the "master catalogue" for each concept.
4. **Blueprints definition.** We design the blueprints according to the needs of the different communities. For each concern they have we propose a blueprint and identify the artifacts types and relations that will be depicted in it. Obviously many concerns are left unanswered due to the lack of information.
5. **Notation definition.** We define a symbol for each artifact and/or relation. Whenever we have encountered a similar concept in UML, we have adopted the UML symbol for such a concept.
6. **Information and Processes Improvements.** We propose changes to IT projects, IT Architecture and IT Governance processes, and define the contents of the documents handled amongst these processes. Process optimization results mainly from a better communication and information flow amongst these different communities. Two important documents are related with the project initiation and conclusion:

   o The way project plan can be processed to identify project dead and alive lists. This can be done by making sure project plans use the proper artifacts names or IDs and uses know keywords to identify if the artifacts are being used, created, updated or deleted within that project.

      o   The description of how a project should present the architecture of the things it is intended to develop, when project starts, or the thing it has developed, when the project ends. This summarizes results of phases 3, 4 and 5.

7. **Automation.** We use our Blueprint Management System (BMS) to gather information from different sources and to produce blueprints and other documentation exchanged between the different processes. The BMS generates office documents automatically, simplifying communication between people.

6 The Blueprint Management System

We now present our Blueprint Management System (BMS), a software solution that implements our approach for generation and maintenance of blueprints.

The BMS collects information from different sources:

- Project Management systems, where information about existing projects and corresponding IDs, dates, resources involved and the artifacts CRUD lists[7], if available in project plans. For example, for each application component related to a project, the following information should be provided: "component; application; layer; platform; CRUD", where the CRUD indicates the action of the project on that component.

- Imported csv files or via a web user interface, where project teams can upload the above information in a textual form, if not existing elsewhere.

- Operational systems, such active nodes from a CMDB or the active services from a SOA service registry. This is information about the production environment that can be used to complement the information gathered from IT projects or to provide alerts. For example, if a project intends to use a service that it does not exist in the production environment (is not <u>alive</u>), nor in the create list of on-going projects (is not in <u>gestation</u>), a warning is issued. As another example, the BMS can query a development platform and find out that a given artifact was created within a given project area, and issues a warning if that artifact was not in project create list.

For each artifact type, the BMS can act as a master or a slave catalog. In the last case, the BMS collects information from existing catalogs periodically, normally in a daily basis. Based in the new information collected the BMS generates the blueprints that may have changed since the last generation.

The blueprints are created both as images and as interactive objects. Regarding the first, users can upload an office document (a Microsoft PowerPoint, Word or Excel format) with references for the desired blueprints and BMS returns the document with the requested blueprints as images. Regarding the interactive objects, they allow both the navigation between different blueprints, since symbols are in fact links to other blueprints, and a querying mechanism based on the values of artifacts properties, that can change the depiction or color of artifacts matching the query.

---

[7] For the sake of simplicity, the BMS considers that projects have four artifacts lists: Create (alive), Read (used), Update (changed), Deleted (dead).

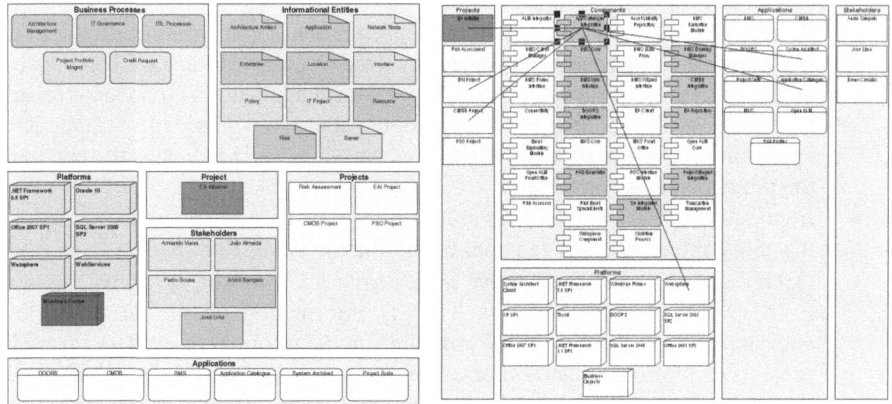

Fig. 2. Governance Blueprint

Fig. 3. Context Blueprint and Project Impact Blueprint

We now present some examples of blueprint generated. The figure 2 presents a Governance Blueprint, produced using the *Organic View* over the type hierarchy {business domain, solution, application}. In this blueprint[8], columns are broad business functions, lines are management level (plan, control, execute) and rectangles inside matrix cells are IT solutions, each aggregating several applications (not depicted). However, the solutions have a color code, according to the delays of on-going projects affecting applications within that solution. One project may affect several applications in several solutions.

The Context Blueprint presented in the left side of figure 3 has a given project as subject and answers two questions: (i) what artifacts affect that project, and (ii) what artifacts are affected by that project. The blueprint presented was produced with $V_{ENV}(\sigma, T_{ENV}, t)$ where $\sigma = Project$ and $T\_ENV = \{Application, ITPlatform, InformationEntity, BusinessProcess, Stakeholder, ITProject\}$ with a color filter according to the CRUD relation between each artifact and the project[9].

---

[8] This blueprint implements the Activity Based Retail Bank Component Map proposed in [1].

[9] The color code may have different semantics for different types of artifacts related.

This blueprints states that project in blue (in the center) intents to remove the platform in red, which is being used by the applications shown in the bottom , and this has an effect on the business processes and information entities, as well as on four other projects.

The details of the dependencies amongst these five projects are detailed in the Project Impact blueprint presented on the right side of the figure 3, where the application components (in UML notation) that are involved in more one project are presented in orange. The image shows the selection of one particular component, and the lines reveal the actual projects where conflict may arise.

The left side of figure 4 presents a Structure Blueprint applied over a given application. This blueprint answers 3 questions: (i) the components of the application; (ii) how they are structured into application layers, and (iii) the platform each component executes. This blueprint is a combination of *Composition* and *Structure Views*. The right side of this Blueprint was produced as $V_{COMP}(\sigma, T_{COMP}, t)$, where

$$\sigma = "Aplication\ and\ "\ T\_COMP = \{UserInterface,\ Integration,$$
$$BusinessLogic, Data\}$$

The left side of this blueprint was generated as $V_{STR}(\sigma, T_{STR}, t)$, where σ is an actual application and

$$T_{STR} = \{(Platform, UserInterface),\ (Platform, Integration),$$
$$(Platform, Business),\ (Platform, Data)\}$$

The right side of figure 4 presents an Integration Blueprint, which depicts how the components are integrated within the organization IT and among themselves. The components appear in UML notation and within the execution platform, and the data flows are links to the typed objects holding the detailed description of the information

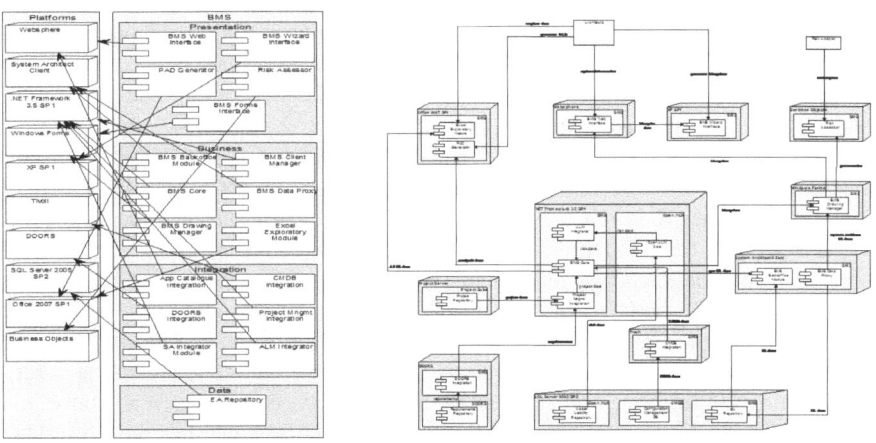

Fig. 4. Structure and Integration Blueprints

and control (push, pull, message, shared database, etc) of that flow. This Blueprint was generated as $V_{STR}(\sigma, T_{STR}, t_i)$, where

$$T_{STR} = \{ \ (IntegrationComponents, Component),$$
$$(Components, \ Platform), \ (Components, Aplication) \ \}$$

and σ is a given application.

The above blueprints are very a small sample of the variety of blueprints that may be produced from the basic views (Organic, Environment, Composition and Structure) described in chapter 4.

The blueprints can be generated either with a static or a dynamic contents. In the last case, a time bar similar to the one presented in figure 5 is added to the top of each blueprint. The time bar has two movable buttons that define a period in time to filter out the displayed artifacts, namely only the artifacts that are in the <u>alive</u> state in that period (01/01/2008 and 05/21/2008 in the example of figure 5) are displayed.

01/01/2008   Effective Date   07/07/2008
                                         01/01/2008 TO 05/21/2008

Fig. 5. The Time Bar

By moving both buttons to the same moment in time, only the artifacts that are in the <u>alive</u> state at that time will be presented. The default position of both buttons is set for today, so the AS_IS situation is presented by default. By moving both buttons forward in time one gets the TO_BE state, as foreseen according to the plans of the projects to be completed before that time.

7   Conclusions

We have successfully managed to have a full set of architectural blueprints being automatically generated on a weekly basis based on information retrieved from IT projects plans, and being used by users of different communities: IT project management, IT Architecture and IT Governance.

The use of blueprints as a common language had several effects in the organization. One unexpected example is the fact that IT project managers changed the way they work. Before, they did project budgeting and planning first and then they think about the architectural representations. Today they do the architecture in the first place, by sending textual data to BMS and getting back the corresponding blueprints, and only then they compile the cost and plan to make each architectural artifact into a global project cost and plan. This answers positively to question Q1 of section 2. Furthermore, it only states that IT projects are doing the same things as what IT architects are architecting and as what IT governance is governing. We believe this is one step forward in the organization self-awareness [20], which has indeed produced by itself unexpected changes.

Regarding question Q2 of section 2, the answer tends to be "yes" when the project starts and ends, and "not as it should" during project execution. This has an impact, especially in long projects, because quite often projects do different things than what they had initially planned, and those changes are not feed into the blueprint analyses and queries done before the end of the project.

Regarding questions Q3 and Q4 of section 2, the answer is definitely "yes". Projects have context, structure and integration blueprints involving only the artifacts relevant to the project, and changes to these blueprints are compiled back to the enterprise wide blueprints.

The answer to question Q5, is simple in theory: as detailed as it needs to be so that everyone can understand the same thing, but in practice one may not achieve a satisfying definition for all artifacts, especially for the non IT domains.

Finally the answer to question Q6 is based on the clarification of the concerns/ stockholders as recommended in [2]. Rather than having a few complex and multipurpose blueprints, one should aim at many blueprints as simple as possible, each answering to one or two simple questions that is useful for a given concern/ stockholder.

We present what we have found in actual companies in banking, telecommunication and retail industries. In some cases, we aim at zero effort blueprints, since they are automatically generated based on information that flows between different communities in the organization, mostly between IT Architecture, and IT Project Management.

So far, we have only experiment our approach in the IT domain and in large organizations. However, the usage of a similar approach in the business domain or even in the IT domain of small organizations may face other difficulties. In fact, our approach requires that artifacts become alive and dead at well known events. In case of IT of large organization, these events occur when artifacts are place into the production environment via well establish procedures. This may not be the case for small companies, where production environment is not so controlled, and for sure, is not the case for business processes domain, where each employee is in fact part of the production environment, making almost impossible to trigger well known events in the organization when business artifacts become alive.

Finally it is worth to say that this work did not evolve as presented here. Even though it had a few cycles of experimentation and formalization, the bulk of formalization presented in chapter 4 did come after the experimentation. We envisage further integration with general systems theory [19] to better address the ideas of control and changeability of organizations as envisaged in [21, 22].

References

[1] IBM, Retail Banking - Business Componet Map. White Paper IBM Global Solution Center (2007), http://www.ibm.com
[2] IEEE Computer Society, IEEE Std 1471-2000: IEEE Recommended Practice for Architecture Description of Software-Intensive Systems. IEEE, New York (2000)
[3] Lankhorst, M.: Enterprise Architecture at Work: Modelling, Communication and Analysis. Springer, Heidelberg (2006)
[4] Pereira, C.M., Sousa, P.: Business Process Modelling through Equivalence of Activity Properties. In: ICEIS 2008, Barcelona, Spain, pp. 137–146 (2008)

[5]   Sousa, P., Caetano, A., Vasconcelos, A., et al.: Enterprise architecture modeling with the UML 2.0. In: Rittgen, P. (ed.) Enterprise Modeling and Computing with UML, pp. 67–94. Idea Group Inc. (2006)

[6]   Zachman, J.: A Framework for Information Systems Architecture. IBM Systems Journal 26(3), 276–292 (1987)

[7]   Open Group. The Open Group Architectural Framework (TOGAF) - Version 8.1 (June 20, 2005), http://www.opengroup.org/togaf/

[8]   Macaulay, A., Mulholland, A.: Architecture and the Integrated Architecture Framework, Capgemini (2006)

[9]   Stader, J.: Results of the Enterprise Project. In: Proceedings of Expert Systems 1996, the 16th Annual Conference of the British Computer Society Specialist Group on Expert Systems, pp. 888–893 (1996)

[10] McFarlan, F.W.: Portfolio approach to information systems. Harvard Business Review, 142–150 (September-October 1981)

[11] Calder, A.: IT Governance: A Pocket Guide. IT Governance Publishing (2007)

[12] Makiya, G.: Integrating Enterprise Architecture and IT Portfolio Management Processes. Journal of Enterprise Architecture 4(1), 27–40 (2008)

[13] Walker, M.: Integration of Enterprise Architecture and Application Portfolio Management (17-02-2009, 2007), http://msdn.microsoft.com/en-us/library/bb896054.aspx

[14] Lagerstrom, R.: Analyzing System Maintainability Using Enterprise Architecture Models. Journal of Enterprise Architecture 3(4), 33–41 (2007)

[15] Op't Land, M., Proper, E., Waage, M., et al.: Enterprise Architecture:Creating value by informed Governance. Springer, Berlin (2009)

[16] Dietz, J.: Enterprise Ontology: Theory and Methodology. Springer, New York (2006)

[17] Reijswoud, V., Dietz, J.: DEMO Modelling Handbook, vol. 1, Version 2.0, Delft, Department of Information Systems, Delft University of Technology, The Netherlands (1999)

[18] Vasconcelos, A., Sousa, P., Tribolet, J.: Information System Architecture Metrics: an Enterprise Engineering Evaluation Approach. Electronic Journal of Information System Evaluation 10 (2007)

[19] Le Moigne, J.-L.: A Teoria do Sistema Geral - Teoria da Modelização (From the original la théorie du système général - Théorie de la modélisation), Instituto Piaget (1997)

[20] Magalhães, R., Zacarias, M., Tribolet, J.: Making Sense of Enterprise Architectures as Tools of Organizational Self- Awareness. Journal of Enterprise Architecture 3(4), 64–72 (2007)

[21] Santos, C., Sousa, P., Ferreira, C., et al.: Conceptual Model for Continuous Organizational Auditing with Real Time Analysis and Modern Control Theory. Journal of Emerging Technologies in Accounting 5, 37–63 (2008)

[22] Matos, M.: Organizational Engineering: An Overview of Current Perspectives, DEI, IST-UTL, Lisbon (2006)

An Information Model Capturing the Managed Evolution of Application Landscapes

Sabine Buckl, Alexander M. Ernst, Florian Matthes, and Christian M. Schweda

Technische Universität München
Boltzmannstr. 3
85748 Garching
{buckls,ernst,matthes,schweda}@in.tum.de

Abstract. Projects are the executors of organizational change and hence in charge of the managed evolution of the application landscape in the context of enterprise architecture (EA) management. Although the aforementioned fact is widely agreed upon, no generally accepted information model addressing the challenges arising in the context of future planning and historization of management decisions concerning projects yet exists. This paper addresses this challenge by identifying requirements regarding an information model for linking projects and application landscape management concepts from an extensive survey, during which the demands from practitioners and the existing tool support for EA management were analyzed. Furthermore, we discuss the shortcomings of existing approaches to temporal landscape management in literature and propose an information model capable of addressing the identified requirements by taking related modeling techniques from nearby disciplines into account.

Keywords: Enterprise Architecture Management, Project Portfolio Management, Temporal Modeling.

1 Introduction

The need for an alignment between business and IT in an organization has been an important topic for both practitioners and researchers ever since the 90's of the last century [15]. Nevertheless, enterprise architecture (EA) management as a means to achieve this alignment has only recently become an important topic, many companies are currently addressing or planning to address in the nearby future. As a consequence of the greater demand from practice, a multitude of approaches to EA management has been proposed in academia [7,19], by standardization bodies [11,26], or practitioners [12,22]. These approaches differ widely concerning the coverage of different aspects of EA management, as e.g. infrastructure or landscape planning. While documentation, analysis, and planning of the EA are common tasks throughout all approaches, the level of abstractness and granularity of information needed to perform EA management differs – for a comprehensive comparison see e.g. [1]. As a consequence, different

A. Albani, J. Barjis, and J.L.G. Dietz (Eds.): CIAO!/EOMAS 2009, LNBIP 34, pp. 85–99, 2009.

information models[1] defining the structure of the respective EA documentation are used in the approaches.

Notwithstanding, certain similarities among the EA management approaches exist, literally all approaches agree on the *application layer* being an important management asset [1,24]. To holistically manage the *application landscape*[2], as the entirety of the business applications and their relationships to each other as well as to other elements, e. g. business processes, of an enterprise, is therefore a widely accepted central task of EA management.

EA management in general and landscape management more specifically can be considered *typical* management processes thus, adhering to the management cycle containing the phases *Plan, Do, Check*, and *Act* [10,25]. Following the periodic phases, the importance of traceability concerning management decisions increases. This means, that the realization of decisions taken in the *Plan* and executed in the *Do* phase is evaluated during the *Check* phase to determine potential process improvements, which are subsequently applied in the *Act* phase. To achieve this type of *self-improving* process, it is necessary to make past decision explicit and accessible during evaluation. The respective technique of archiving management decisions is usually called *historization*. A typical EA management question, which needs historic information for answering, could be: Is the status of a planned application landscape reached within the planned time frame or has the plan been changed? The type of time-dependence employed in making that information available is different from the dependence as alluded to above, such that it has to be incorporated into an information model for EA management separately. From this discussion the following research question guiding the remainder of the article can be derived:

> How should an information model for landscape management be designed to incorporate both business and technical aspects, and to support future planning and historization of management decisions?

This question especially alludes to the aspects of *time-dependency* as connected to application landscape management. Therein, different types of landscapes are of importance [3]:

- the *current* landscape, reflecting the actual landscape state at a given point in time,
- *planned* landscapes, which are derived from planned projects for transforming the landscape until a certain point in time, and
- the *target* landscape, envisioning an ideal landscape state to be pursued.

---

[1] Consistent to the terminology as used e. g. in [8], we call the meta model for EA documentation an *information model*.

[2] We do not use the term *application portfolio*, which is widely used in this area, as we regard it to have a narrower focus on just the business applications without considering related artifacts, as e.g. business processes.

[3] The different types are sometimes also called *as-is* and *to-be* landscapes, thereby abstaining from the distinction between planned and target landscape.

This information has to be maintained consistently in an integrated information model for the EA, describing the three types of landscapes adequately. Therein, especially the relationship between application landscape and project portfolio management, concerning the planned landscapes, has to be considered. To ensure consistency in between these management tasks, the planned landscape should be derived from the transforming projects selected for execution in the next planning cycle. The main purpose of the project portfolio management process is to identify those projects, which should be accomplished [29]. Thereby, a distinction has to be made between projects that have to be performed, e.g. due to end-of-support – *run the enterprise* – and projects that should be performed, e.g. due to strategic reasons – *change the enterprise* [3].

The close linkage between application landscape and project portfolio management should be used to create different planned landscapes for the next planning period based on distinct project portfolio selections. Thereby, analyses of the application landscape can be used to provide decision support for project portfolio management. In order to identify the appropriate portfolio, analyses regarding the dependencies between projects have to be conducted. Therefore, not only the required resources, e.g. persons, tools, etc, need to be considered but also dependencies between projects regarding affected artifacts, e.g. business applications, interconnections, etc. These analyses support the identification of potential conflicts regarding time-dependencies if a project in the realization phase is delayed. The aspect of time, as related to projects, must be considered a highly complex one, which is also not well reflected in current information models for landscape management (cf. Section 2 and [6]).

For the planned and target landscape different (historic) states may exist and evolve over time, i. e. a planned landscape is planned for a certain point in time and is modeled at a certain (different) point in time. The latter is also true for the target landscape, which might be modeled differently at different points in time. Additionally, the idea of *variants* for planned landscapes resulting from the selection of different project portfolios has to be taken into account. Summarizing, three different *time-related* dimensions exist:

- firstly, a landscape is *planned for* for a specific time[4],
- secondly, a landscape is *modeled at* a certain point in time, and
- thirdly, different landscape *variants* of a planned landscape may exist.

Figure 1 illustrates the relationships between current, planned, and target landscapes as well as the different dimensions relevant for landscape management.

In the remainder of the article, the research question, as alluded to above, is approached as follows: Section 2 gives an overview about current approaches to landscape management as described by researchers and practitioners in this field. In Section 3, requirements – especially time-related ones – for an information model for landscape management are elicited. These requirements are subsequently incorporated in a temporal information model for documenting and

---

[4] Therein, the current landscape can be regarded to be planned for the current time, while the target landscape can be regarded to be planned for an infinite future point in time.

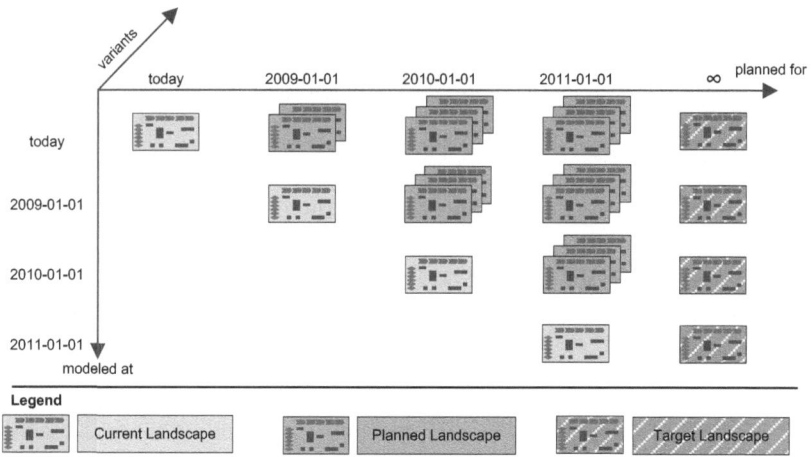

Fig. 1. Current, planned, and target landscape

planning application landscapes, presented in Section 4. Final Section 5 concludes the paper and briefly sketches interesting directions for future research.

2 Related Work

Application landscape management is, as outlined before, widely considered an important task in EA management. Consequently, a number of different ways to approach this task have been proposed both in practice and academia. Subsequently, we briefly introduce selected approaches and especially emphasize on the coverage of time-related aspects.

The EA management approach presented by van der Torre et al. in [27,28] considers the business applications as well as their relationships to other elements of the EA an important information asset. As part of the approach, ways to display this information to managers in an appropriate way to provide decision support are discussed. The recommended type of visualization introduced in the paper, is the so-called *landscape map* detailing the *business applications* in relationship to the provided *business functions* and the respective *products* of the company. This relationship is referred to as a ternary one. Aspects of time-dependency are not discussed, also projects are not alluded to in the article, as the approach mainly focuses on application landscapes.

In [4], Braun and Winter present the application landscape as the set of an enterprise's business applications and their interdependencies. Respectively, the information model contains the classes *application* and *interface* to describe the web of dependencies. These concepts from the *application layer* can be connected to elements from the *organizational layer*, i.e. the *business processes* in order to explicate the business support relationships. Nevertheless, the model does not account for analyses in multi-organizational unit environments, as no concept for relating the organizational unit, where the business support takes place, is provided. Further, the concept of time-dependence is only partially addressed in the information model – means for planning landscape transitions via projects are not included therein.

The interfaces connecting the business applications are also focused by Garg et al. in their approach presented in [14]. Therein, especially the total number of interfaces associated with an application is regarded an important quantitative indicator, which should be taken into consideration, e.g. if changes are applied to the application landscape. Consequently, the approach emphasizes on analyses regarding the current landscape, while planned landscapes in special and time-related concepts in general are not part of the approach. Additionally, due to the strong application focus of the model, business aspects, i.e. business processes are omitted in the model. Also, transformation projects changing the application landscape are not alluded to in the approach.

As part of the *systemic enterprise architecture methodology (SEAM)* Le and Wegmann [20] discuss a way to model enterprise architectures for reasons of documentation and analysis. Prior to presenting the model, some requirements are introduced, especially focusing on *multi-level-modeling* from company level down to component level. Furthermore, the importance of traceability is emphasized, although not targeting temporal traceability but inter-level traceability of relationships. The approach further introduces the more abstract concept of the *computational object*, effectively replacing the business application. Nevertheless, time-related aspects are not discussed in the approach; projects as executors of organizational change, which correspond to the external drivers such as new competitors, laws, changing markets, etc., are also not alluded to.

Jonkers et al. present in [17] a language for enterprise modeling, in which they target the three layers of *business*, *application*, and *technology*. The concepts introduced on the different layers can be used for modeling the current application landscape, especially for explicating the business support provided by applications (components) via offered interfaces. Further, the approach refines the description of the business support by adding the supplemental concepts of *business-* and *application-services* respectively. These concepts can be used to describe the existence of a support without having to specify, which actual application is responsible for the support. Thereby, target landscape planning could be facilitated. Nevertheless, planned landscapes are not in the scope of the model, which also contains no concept for modeling projects or explicating project dependencies.

The approach of *multi-perspective enterprise modelling (MEMO)* as discussed e.g. in [13] explicitly accounts for the modeling of IT concepts, as business applications, in an organizational and business context, described as *organizational units* and *roles* as well as *business processes* and *services*. The respective modeling language concerned with IT aspects is the *IT modeling language (ITML)* [18] introduces the respective concepts, as e.g. the *information system*. According to the reference process described as complementing the language, these concepts should not only be used for documentation, but also for landscape planning.

Nevertheless, projects are not part of the model, which also does not explicitly account for issues of time-dependence.

Beside the academic community, also practitioners address the field of EA management and landscape management. A representative approach developed by a consulting company is the *QUASAR Enterprise* approach [12]. In this approach the application landscape is presented as management subject related with business and technical concepts, ranging from *business processes* to *technical platform modules*. The current landscape is consequently documented with references to these concepts. Complementing, a so-called *ideal landscape*[5] should be defined in the application landscape management process. Different *to-be* (planned) landscapes are created describing the transition roadmap from current towards target landscape. These intermediary roadmap landscapes should according to the approach maintain relationships to the respective projects, nevertheless means for tracing back the evolution of the planned landscapes are not discussed in the approach.

The Open Group is a consortium of practitioners addressing the field of EA management, whose main purpose is to develop standards in this area. Therefore, they proposed *The Open Group Architecture Framework (TOGAF)* [26], which provides a cyclic process model for EA management, the *Architecture Development Method*. This cycle contains, among others the phase *architecture vision*, which is concerned with the development of an target architecture. In order to manage an evolution in the direction of the target architecture, several intermediate *transition architectures*, which we would refer to as planned landscapes, are developed. In addition to the cyclic process, TOGAF 9 [26] includes an information model, which describes the content of architecture development. Thereby, projects are introduced via the concept of *work packages*. Nevertheless, these work packages are neither linked in the information model to the architectural elements, which they affect, nor does the provided information model provides concepts to model time-dependencies between the different elements.

3   Elicit Requirements for Landscape Management

Due to great interest from industry partners in information about EA management tools and especially their capabilities to address the concerns arising in the context of landscape management, an extensive survey – the *Enterprise Architecture Management Tool Survey 2008* – was conducted in 2008 [21]. The survey pursues a threefold evaluation approach, relying on two distinct sets of scenarios together with an online questionnaire. The survey was developed in cooperation with 30 industry partners (among others Allianz Group IT, sd&m – software design & management, Siemens IT Solutions and Services, Munich Re, O2 Germany, BMW Group, Nokia Siemens Networks). Thereby, the first set of scenarios focuses on specific functionality, an EA management tool should provide, without connecting these functionalities to the execution of a typical EA management task, e.g. *1) flexibility of the information model, 2) creating*

---

[5] *Target* landscape in the terms used throughout this paper.

visualizations, or *3) impact analysis and reporting*. The EA management tools are further evaluated by the scenarios of the second set, which reflect tasks that have been identified as essential constituents of many EA management endeavors, e.g. *1) business object management, 2) IT architecture management*, or *3) SOA transformation management*. One of the most prominent scenarios of the second part is the scenario *landscape management*, which is concerned with the managed evolution of the application landscape [2]. The concern of the scenario was described by the industry partners as follows:

Information about the application landscape should be stored in a tool. Starting with the information about the current landscape potential development variants should be modeled. The information about the current application landscape and future states should be historicized to enable comparisons. [21]

Closely related to the landscape management scenario is the *project portfolio management*, which is concerned with providing decision support for the selection of an appropriate portfolio of projects to be realized within the next planning period as alluded to in Section 1. Subsequently, a catalog of typical questions in the context of landscape and project portfolio management as raised by the industry partners is given:

– What does the current application landscape look like today? Which business applications currently support which business process at which organizational unit?
– How is, according to the current plan, the application landscape going to look like in January 2010? Which future support providers support which business process at which organizational unit?
– What was, according to the plan of 01-01-2008, the application landscape going to look like in January 2010?
– How does the target application landscape look like?
– What are the differences between the current landscape and the planned landscape, according to the current plan? What are the differences' reasons?
– What are the differences between the planned landscape according to the plan of 01-01-2008 and the current plan?
– What projects have to be initiated in order to change from the planned landscape (according to the current plan) to the target landscape? What planning scenarios can be envisioned and how do they look like?
– Which EA artifacts, e.g. business applications, are modified/created/retired by the individual project proposal?
– Which project proposals (run the enterprise) have to be accomplished in any case?

Based on the questions from the industry partners, the different landscape types, and time-related dimensions relevant for landscape management (see Section 1), the following requirements regarding an information model can be derived. An information model suitable for landscape management must:

R1 contain a ternary relationship in order to support analyses regarding current and future business support (which business processes are supported by which business applications at which organizational units),

R2 provide the possibility to specify envisioned business support providers in order to facilitate target landscape planning without having to specify implementation details of the business support,

R3 support the deduction of future landscapes from the project tasks, which execute the transition from the current to the future business support,

R4 ensure the traceability of management decisions by storing historic information of past planning states, which may be interesting especially if complemented with information on the rationale for the decisions,

R5 foster the creation of landscape variants based on distinct project portfolios in order to tightly integrate project portfolio management activities, and

R6 allow impact analyses regarding dependencies between different projects, which affect the same EA elements, e.g. organizational units, business application, business processes.

These requirements have been used in [6] to evaluate the support for landscape management as provided in approaches from literature and in approaches provided in three commercially available EA management tools. The result of this evaluation is that none of the analyzed approaches completely fulfills all requirements given above.

4 Developing a Temporal Information Model

In this section, we present an information model capable to fulfill the requirements as introduced above. Hence, the model addresses the research question as stated in Section 1. Such a model could be described using different modeling languages, of which an object-oriented one – namely the UML – has been chosen. This choice seems to us equally suitable to potential alternatives, as e.g. Entity/Relationship (E/R) modeling. This opinion is supported by the fact that the subsequently presented object-oriented information model can be easily converted to an E/R model. It has further to be noted that we do not regard the UML as the language of choice for presenting the information modeled according to the information model – other graphical notations, i.e. means to define viewpoints [16], exist, which a by far more appropriate for visualizing enterprise architecture information. The object-oriented information model hence only defines the schema for storing this information.

To prepare the discussions on the information model, we provide a short glossary (see Section 4.1) of core model concepts. These concepts are reflected in the information provided in Section 4.2, which is an augmentation of a model initially discussed in [6].

4.1 Glossary

In this section, the core concepts relevant in application landscape management are introduced and defined in an informal way. The definition are taken from the glossary as presented in [7], although minor adaptations have been applied to suite the specific setting of the article.

Business application. A business application refers to an actual deployment of a software system in a certain version at a distinct location and hardware. Thus, business applications maintain a versioning information in addition to the relationships to the business processes, they support at specific organizational units. In landscape management, the business applications are limited to those software systems, which support at least one business process. Further, the business applications are the objects, which are transformed by the projects considered in application landscape management.

Business process. A business process is defined as a sequence of logical, individual functions with connections in between. A process here should not be identified with a single process step, as found e.g. in an *event driven process chain (EPC)*. It should be considered a coarse grained process at a level similar to the one used in value chains, i.e. partially ordered, linear sequences of processes. Additionally, a process maintains relationships to the business applications, which support him at the different organizational units.

Envisioned support provider. An envisioned support provider is a constituent of a target application landscape, used to indicate that a related business process is supported at a distinct organizational unit, without giving a specification, which business application is likely to provide this support, if any. Inspite of the similarities to the business application, the envisioned support provider is not affected by projects but has nevertheless a period of validity associated. Thereby, it references the point in time it has been modeled at and (optional) the point in time, the envisioned provider became invalid.

Organizational unit. An organizational unit represents a subdivision of the organization according to its internal structure. An organizational unit is a node of a hierarchical organization structure, e.g. a department or a branch.

Project. Projects are executors of organizational change. Therefore, adaptations of the application landscape are the result of a project being completed. Projects are scheduled activities and thus hold different types of temporal attributes, their *startDate* and *endDate* on the one hand. On the other hand, projects are *plannedAt* respectively *removedAt* certain points in time referring to the time of their creation or deletion. This effectively results in a period of validity, which is assigned to each project. In application landscape management, projects are considered to only affect business applications in general and their business support provided, in special. Projects can be split into smaller constituents, so called *project tasks*.

With these core concepts and main attributes at hand, an information model satisfying the requirements corresponding to application landscape management can be developed.

4.2 An EA Information Model for Modeling Project Dependencies

Based on the discussions in [6] an information model has been proposed. This information model (cf. Figure 2) is capable to satisfy the requirements **(R1)** to **(R4)**. It also fulfills **(R5)** to a certain extent. Landscape variants, based on certain project selections, i.e. planned project portfolios, can be derived from the model at any point in time. Nevertheless, these variants are not historized, as the model does not contain a concept for storing different portfolio selections. We do not regard this a major issue, because the project selections are most commonly used in a project portfolio management discussion process, which leads to a certain selection to be approved. Additionally, making it possible to store different selections or, even more sophisticated, different timelines for the projects in a long-term project planning would introduce a number of additional concepts. This seems to us especially cumbersome, as the consequential complexity in creating model instances, might not relate to the benefits earned from this additional instrument of future planning. Furthermore, the practitioners, which have raised the requirements **(R1-R5)** (cf. [21]), did not state such medium-term multi-project portfolio variants as a topic of interest.

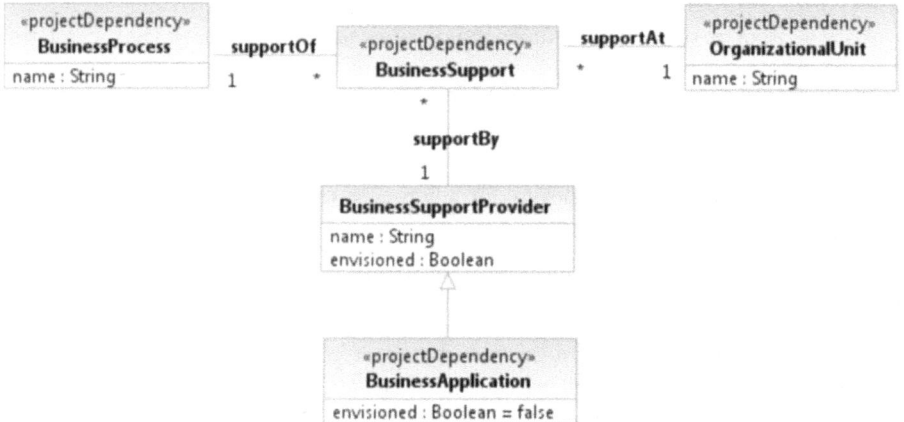

Fig. 2. Information model satisfying **(R1)**, **(R2)**, **(R3)**, **(R4)**, and **(R5)**

The aforementioned model uses two UML stereotypes [23] (`<<temporal>>` and `<<projectDependency>>`) to enhance model clarity and support concise modeling. As these stereotypes cannot be considered widely-known, they are subsequently explained in detail.

The first stereotype `<<temporal>>` has been proposed in [9] in the context of the modeling pattern *temporal property*[6]. This pattern allows to model that a property of an object can change over time and these changes have to be tracked. Nevertheless, using this pattern to address issues of time-dependency for properties does not come without costs – the attribute, which is converted to a temporal property, is changed to a multi-valued one, i.e. one of multiplicity ∗. A class owner may have exactly one value for a property assigned at a specific point in time. Nevertheless, there may be multiple instances of the respective value

---

[6] This pattern is also known as *historical mapping* or *time-value pairs*.

class assigned to the same owner, as they represent the history of property values over time. This issue is resolved by introducing the <<temporal>> stereotype, indicating that a property might have multiple values without overlap in their periods of validity.

The second stereotype <<projectDependency>> is introduced, to support concise modeling of the relationship between the projects, i.e. their constituing tasks, and the architectural constituents. A project task can affect an architectural constituent in four different ways:

- *Introduce* a new constituent to the architecture.
- *Migrate* from one constituent to another, e.g. a functionality.
- *Retire* a constituent from the architecture.
- *Change* an existing constituent *below* the EA level.

The first three types are quite obvious, although the fourth type is also important, as it can be used to subsume changes *below* the architectural level. These may e.g. be changes to the components of a business application leading to a new version of the application, although the application itself has not changed from an EA point of view. Making projects performing minor changes explicit is necessary to completely fulfill (**R6**), in order to prevent multiple projects from performing concurrent and potentially conflicting changes.

EA constituents, which can be affected by projects, must hence be related to the corresponding project tasks, which can be achieved in many different ways in the information model. A maximum of genericity can be reached by introducing a basic concept for any concept, which can be affected by a project or a part thereof and to use respective inheritance in the information model. We further pursue this approach and introduce the respective basic concept and its associations to *project tasks*, which are used to model distinct activities within a project. The model incorporating this idea is shown in Figure 3.

In this information model, any *project affectable* can derive its period of validity from the *start* and *end dates* of the transitively associated projects. Thereby, inheriting from *project affectable* makes it possible to assign a project dependency to a concept in the information model. Nevertheless, using the standard UML-notation for inheritance would make the model less easy to perceive, as many classes are likely to inherit from *project affectable*. To make the resulting model more concise, we introduce an additional stereotype <<projectDependency>>, which can be assigned to a class in order to indicate, that this class is actually a subclass of *project affectable*.

In order to ensure model consistency, a modeling constraint applies – defining that a project task might not migrate between EA constituents of different types:

```
inv: Migration
introduces.type == retires.type
```

Completing the information model the value of the derived attribute *isMaintenace* is complemented with a computation formalism to automatically derive

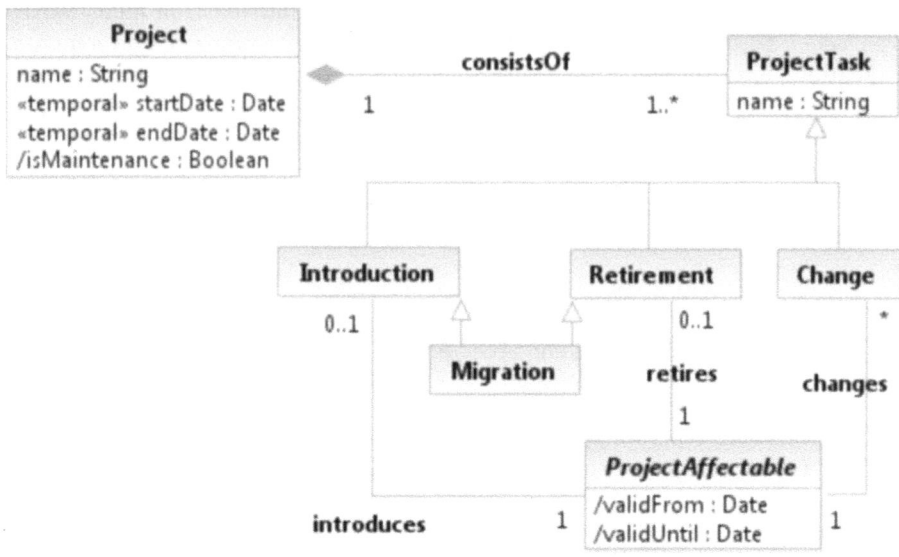

Fig. 3. Project affectable and project with exemplary child class

the distinction between *maintenance* and *transformation* projects as discussed above.

```
derive: Project
isMaintenance = consistsOf->forAll(t|t.oclIsTypeOf(Change))
```

The periods of validity for an architectural constituent are derived, as alluded above, from the associated project tasks:

```
derive:ProjectAffectable
validFrom = introduces==null?null:introduces.project.endDate
```

A similar derivation rules applies for the end date for the period of validity. For both dates, the special value *null* can be computed, which indicates that the corresponding architectural element has no distinct date of introduction or retirement. This means that the project, which introduced the element, took place before EA documentation was introduced or that no retirement project is yet planned respectively. For further discussions on how to incorporate these special dates into landscape transformation planning see e.g. [5].

5   Reflection and Outlook

In this article, we discussed techniques for modeling the project dependencies of EAs in general and application landscapes more specifically. In Section 2 we considered selected state of the art approaches to EA management, having a special emphasis on their support for explicating project- and time-dependencies in their information models. Requirements for modeling the management evolution of an application landscape, which have been collected at EA management practitioners (cf. [21]), were outlined in Section 3. Subsequently, we created and presented an information model capable of fulfilling these requirements. Therein, we applied temporal patterns, i.e. patterns for things that change over time.

The information model presented in this paper has yet not been validated in practice. In doing so especially the complexity of the project dependency modeling might be a usage impediment, which should be addressed by an appropriate user interface. Such an interface can be helpful to conceal large parts of the complexity – thereby making a convenient modeling experience possible. Nevertheless, no such user interface has yet been created, which would be a prerequisite to testing the information model in a practical environment.

The model introduced in the paper is further limited to projects affecting business applications, business processes, organizational units, and their relationships to each other. This does not completely reflect on the role of the project in EA management in general, as a project can also affect and change other EA constituents, such as e.g. infrastructure components or hardware devices. The concept of the *project affectable* as presented in Section 4 could nevertheless be extended to other EA constituents and hence form a reusable building block for incorporating project dependencies in EA information models. This relates well to the approach of EA management patterns as presented in [7], although more in-depth research is yet to be undertaken.

The latter discussion points towards another interesting direction of research. Object-oriented modeling languages, albeit their wide proliferation as discussed in Section 4, do not provide dedicated means for constructing time- and project-dependent EA information models. Hence, techniques as temporal patterns have to be utilized. These techniques could nevertheless by incorporated in an augmented object-oriented modeling language with specific support for creating EA information models. Future research is to show, how such a language could look alike.

References

1. Aier, S., Riege, C., Winter, R.: Unternehmensarchitektur – Literaturüberblick Stand der Praxis. Wirtschaftsinformatik 50(4), 292–304 (2008)
2. Aier, S., Schönherr, M.: Enterprise Application Integration – Flexibilisierung komplexer Unternehmensarchitekturen, Gito, Berlin (2007) (in German)
3. Aier, S., Schönherr, M.: Flexibilisierung von Organisations- und IT-Architekturen durch EAI. In: Enterprise Application Integration – Flexibilisierung komplexer Unternehmensarchitekturen Band I, Berlin, Gito (2007) (in German)
4. Braun, C., Winter, R.: A comprehensive Enterprise Architecture Metamodel. In: Desel, J., Frank, U. (eds.) Enterprise Modelling and Information Systems Architectures 2005. LNI, vol. 75, pp. 64–79. GI (2005)
5. Buckl, S., Ernst, A., Kopper, H., Marliani, R., Matthes, F., Petschownik, P., Schweda, C.M.: EAM Pattern for Consolidations after Mergers. In: SE 2009 – Workshopband, Kaiserslautern (2009)

6. Buckl, S., Ernst, A., Matthes, F., Schweda, C.M.: An Information Model for Landscape Management – Discussing temporality Aspects. In: Johnson, P., Schelp, J., Aier, S. (eds.) Proceedings of the 3rd International Workshop on Trends in Enterprise Architecture Research, Sydney, Australia (2008)

7. Buckl, S., Ernst, A.M., Lankes, J., Matthes, F.: Enterprise Architecture Management Pattern Catalog, Version 1.0. Technical report, Chair for Informatics 19 (sebis), Technische Universität München, Munich (February 2008)

8. Buckl, S., Ernst, A.M., Lankes, J., Matthes, F., Schweda, C., Wittenburg, A.: Generating Visualizations of Enterprise Architectures using Model Transformation (Extended Version). Enterprise Modelling and Information Systems Architectures – An International Journal 2(2) (2007)

9. Carlson, A., Estepp, S., Fowler, M.: Temporal patterns. In: Pattern Languages of Program Design. Addison Wesley, Boston (1999)

10. Deming, E.W.: Out of the crisis, Massachusetts Institute of Technology, Cambridge (1982)

11. Department of Defense (DoD) USA. DoD Architecture Framework Version 1.5: Volume I: Definitions and Guidelines (cited 2008-03-19) (2008), http://www.defenselink.mil/cio-nii/docs/DoDAF_Volume_I.pdf

12. Engels, G., Hess, A., Humm, B., Juwig, O., Lohmann, M., Richter, J.-P.: Quasar Enterprise – Anwendungslandschaften serviceorientiert gestalten. dpunkt.verlag, Heidelberg (2008)

13. Frank, U.: Multi-perspective Enterprise Modeling (MEMO) – Conceptual Framework and Modeling Languages. In: Proceedings of the 35th Annual Hawaii International Conference on System Sciences, vol. 35, pp. 1258–1267 (2002)

14. Garg, A., Kazman, R., Chen, H.-M.: Interface Descriptions for Enterprise Architecture. Science of Computer Programming 61(1), 4–15 (2006)

15. Henderson, J.C., Venkatraman, N.: Strategic Alignment: Leveraging Information Technology for Transforming Organizations. IBM Systems Journal 38(2-3), 472–484 (1999)

16. IEEE. IEEE Std 1471-2000 for recommended Practice for Architectural Description of Software-intensive Systems (2000)

17. Jonkers, H., Goenewegen, L., Bonsangue, M., van Buuren, R.: A Language for Enterprise Modelling. In: Lankhorst, M. (ed.) Enterprise Architecture at Work. Springer, Heidelberg (2005)

18. Kirchner, L.: Eine Methode zur Unterstützung des IT-Managements im Rahmen der Unternehmensmodellierung. PhD thesis, Universität Duisburg-Essen, Logos, Berlin (2008)

19. Lankhorst, M.: Enterprise Architecture at Work: Modelling, Communication and Analysis. Springer, Heidelberg (2005)

20. Le, L.-S., Wegmann, A.: Definition of an Object-oriented Modeling Language for Enterprise Architecture. In: Proceedings of the 38th Annual Hawaii International Conference on System Sciences, 2005. HICSS 2005, p. 179c (2005)

21. Matthes, F., Buckl, S., Leitel, J., Schweda, C.M.: Enterprise Architecture Management Tool Survey 2008. Chair for Informatics 19 (sebis), Technische Universität München, Munich (2008)

22. Niemann, K.D.: From Enterprise Architecture to IT Governance – Elements of Effective IT Management. Vieweg+Teubner (2006)

23. OMG. Unified Modeling Language: Superstructure, Version 2.0, formal/05-07-04 (2005)

24. Pulkkinen, M.: Systemic Management of Architectural Decisions in Enterprise Architecture Planning. Four Dimensions and three Abstraction Levels. In: Proceedings of the 39th Annual Hawaii International Conference on System Sciences, 2006. HICSS 2006, vol. 8, p. 179c (2006)
25. Shewart, W.A.: Statistical Method from the Viewpoint of Quality Control. Dover Publication, New York (1986)
26. The Open Group. TOFAF Enterprise Edition Version 9 (2009)
27. van der Torre, L.W.N., Lankhorst, M.M., ter Doest, H.W.L., Campschroer, J.T.P., Arbab, F.: Landscape Maps for Enterprise Architectures. Technical report, Information Centre of Telematica Instituut, Enschede, Netherlands (2004)
28. van der Torre, L.W.N., Lankhorst, M.M., ter Doest, H.W.L., Campschroer, J.T.P., Arbab, F.: Landscape Maps for Enterprise Architectures. In: Dubois, E., Pohl, K. (eds.) CAiSE 2006. LNCS, vol. 4001, pp. 351–366. Springer, Heidelberg (2006)
29. Wittenburg, A.: Softwarekartographie: Modelle und Methoden zur systematischen Visualisierung von Anwendungslandschaften. PhD thesis, Fakultät für Informatik, Technische Universität München (2007)

A Service Specification Framework for Developing Component-Based Software: A Case Study at the Port of Rotterdam

Linda Terlouw[1] and Kees Eveleens Maarse[2]

[1] Delft University of Technology,
Mekelweg 4, 2628 CD Delft, The Netherlands
l.i.terlouw@tudelft.nl
[2] Port of Rotterdam,
Wilhelminakade 909, Rotterdam, The Netherlands
cs.eveleens.maarse@portofrotterdam.com

Abstract. This paper describes the results of a case study conducted at the Port of Rotterdam, the largest port in Europe. The goal of the case study is to evaluate our Enterprise Ontology-based service specification framework for its use in component-based software development projects. During a Rational Unified Process (RUP)-project at the Port of Rotterdam we specified the required services for the first iterations using this framework. The framework contributed to early error discovery and awareness of important, but often overlooked, service aspects. Overall the service specification framework fulfilled the needs of the project, though some findings led to improvements in our framework.

Keywords: Service Specification, Component-Based Development, Design Science Research, Case Study.

1 Introduction

Structuring large, complex software systems is not an easy task to master. In the '70s Parnas started out with answering the question on how to decompose systems into modules [16]. During the '90s the focus shifted towards finding the right objects [18,4,13]. Nowadays, we are dealing with identifying the right components [14,3] and services [15,24]. The benefits of structuring a software system, whether it is in modules, objects, or components, still remain the same, i.e. (i) making the total software system structure more comprehensible, (ii) enabling easy replacement of components of the software system, (iii) making it possible to divide work between several groups of developers without them needing to be aware of the structure of the total software system, and (iv) minimizing the effect that changes in one part of the system have on the other parts of the system.

Though academic theory provides us with criteria for constructing software system of smaller parts, it is often hard to apply these criteria in practice. One

A. Albani, J. Barjis, and J.L.G. Dietz (Eds.): CIAO!/EOMAS 2009, LNBIP 34, pp. 100–114, 2009.

has to take into account legacy systems written in all kinds of programming languages following different paradigms. Also, the 'reuse before buy before build' principle becomes more and more popular. This principle results in hybrid software systems consisting of legacy systems, COTS components, and newly built components. Since software architects and developers have no control over COTS components, and often not enough knowledge available on legacy systems, they can only apply the component identification principles to a certain extent.

This paper describes a case study conducted at the Port of Rotterdam in which we are dealing with such a hybrid software system. Our subject of study is a service specification framework that can be used as a template for specifying the services through which the components interact. The goal of these service specifications is to allow the developers of different components to work independently during the software development project; all the information they require for the interaction with other components is documented in the service specifications.

The paper continues with our research methodology in section 2. We present our service specification framework in section 3. In section 4 we introduce the Port of Rotterdam and the HaMIS project. We have applied the framework during the first iterations[1] of this project. We discuss the evaluation of our service specification framework in section 5. In section 6 we conclude by summarizing the main contributions of the framework in the project and the findings that can lead to improvements of the framework.

2   Research Methodology

The goal of *Information Systems* (IS) research is to produce knowledge that enables the application of information technology for managerial and organizational purposes [11]. Looking at the IS research field, we see a distinction between (explanatory) *behavioral science* research and *design science* research. In our case we are, speaking with the terminology of Van Strien [21], not so much interested in an 'explanandum' as research object and the creation of a causal model as the behavioral sciences usually are. Instead, we are interested in a 'mutandum' as research object and the creation of a prescription as research product. Therefore, we classify this research into the design science research paradigm [12,20].

The most widely accepted framework for design science research is that of Hevner et al. [12]. However, it lacks an explicit process description. For this reason we have applied the *Design Science Research Methodology* (DSRM) of Peffers et al. [17]. The DSRM, depicted in Fig. 1, is based on seven papers about design science research including the paper of Hevner et al. In the remainder of this section we explain how we followed each process step. We took a problem-centered initiation as our research entry point and followed the nominal sequence, starting with the first activity: "identify problem and motivate". In Fig. 2 we

---

[1] Though we are well aware of the difference between incremental and iterative development, we use the term 'iterative' in this paper for the combination of both since RUP does not distinguish between them.

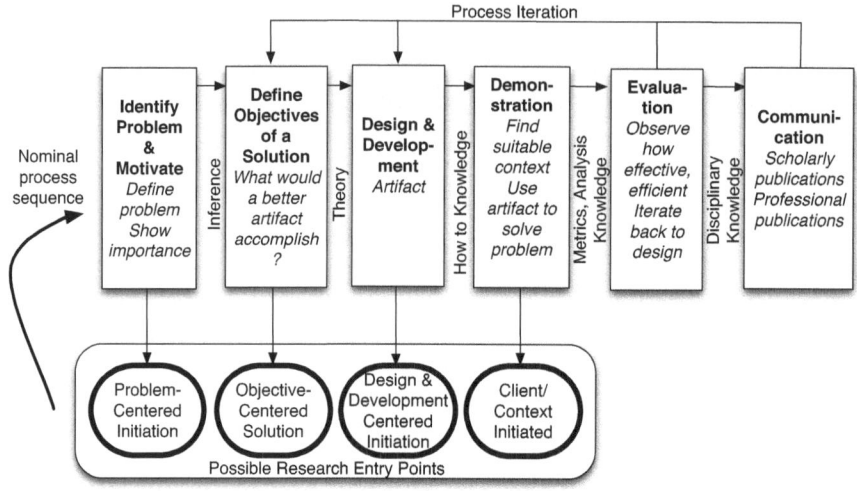

Fig. 1. DSRM Process Model [17]

summarize how adopting this process led to conforming to the guidelines of Hevner et al.

Identify Problem and Motivate. The problem we have identified is as follows. To construct a component-based software system, we need to specify the services offered by each component in a uniform way providing enough information to the people responsible for the other components. An important aspect for documenting software interfaces of any kind is to be as specific and precise as possible because an interface specification that two different parties can interpret differently is likely to cause problems and confusion [5]. Standards from the Web services stack are currently the most widely used standards for realizing services for enabling (inter)organizational reuse of IT functionality, but also for standardizing the connections between components of complex software systems. No standardized way for specifying the whole service behavior (and not only the interface) exists.

Define Objectives of a Solution. Our main research question is: "Of what aspects does a service specification framework consist?". The objective of our framework is to support people in making complete and precise service specifications. This serves multiple purposes, viz. enabling parallel work by provider and consumer, early error discovery, finding services, and making each other's responsibilities clear upfront. Current approaches we have encountered, aiming at solving the same problems, are: the UDDI, several Semantic Web approaches like OWL-S, WSMO, and WSDL-S, and the business component specification framework of Ackermann et al. [1].

Design and Development. The artifact of our research is a service specification framework that can be used as a template for specifying services in component-based software development projects. For creating our artifact, being a service specification framework, we take the Enterprise Ontology, underlain by the Ψ theory, as a starting point. This Ψ theory finds it roots in the scientific field of Language Action Perspective (LAP) [9,8,10].

Demonstration. Before starting with the case study for thoroughly evaluating the framework, we have demonstrated the framework using two service examples from the life insurance industry, viz. the CalculatePremium service and the RegisterAdvice service. We have modeled the complete Enterprise Ontology of this life insurance company.

Evaluation. A case study at the Port of Rotterdam that is structured according to the ideas of Yin [23] acts as a means of evaluation. A case study is an empirical inquiry that (i) investigates a contemporary phenomenon within its real-life context, especially when (ii) the boundaries between phenomenon and context are not clearly evident [23]. A study can involve a single or multiple cases and may use qualitative and quantitative data.

Communication. Currently we are working on several scientific and professional publications. Also, the work has been presented at the Port Of Rotterdam, Ordina, a Dutch governmental organization, a large organization in the aviation sector, a Dutch life insurance/pension company (on which the demonstration was based), an international leasing company, and an international bank.

---

1 *Design as an Artifact:* Our artifact is a service specification framework

2 *Problem Relevance:* Thorough service specifications are required to make service provider and consumer successfully interact. If only the interface (in terms of input, output and errors) is specified this can lead to serious misinterpretations on what the service actually does.

3 *Design Evaluation:* We have used an observational design evaluation method, i.e. a case study.

4 *Research Contributions:* The contribution of this research is a standardized way of specifying services.

5 *Research Rigor:* The framework is based on the Enterprise Ontology, underlain by the Ψ theory [6]

6 *Design as Search Process:* The framework is currently applied and evaluated in a software development project. In future research we plan to apply it for other purposes, i.e. for (IT) service discovery and for human services specification.

7 *Communication:* The artifact has been presented to six large organizations. Scientific and professional publications are in progress.

Fig. 2. Conforming to Guidelines of Hevner et al.

3   The Service Specification Framework

3.1   Theoretical Background

As we mentioned in section 2 we have based our framework on the notion of Enterprise Ontology [6]. The Ψ theory that underlies this notion of Enterprise Ontology finds it roots in the scientific field of Language Action Perspective (LAP). The Ψ theory, consisting of four axioms and one theorem, regards an enterprise as a purposefully designed and engineered social system. This means that the theory focuses on the communication between social actors, being human beings, for modeling enterprises. Though we founded our service specification framework on the notion of enterprise ontology, it can also be used in situations in which the business models and the services are not derived from the enterprise ontology theory (which is the case at the Port of Rotterdam). The details on how we derived the framework from the Enterprise Ontology framework are beyond the scope of this paper and will be subject of another paper. The current scope is the evaluation of the framework in the case study.

3.2   Explanation of the Framework

Figure 3 depicts our service specification framework. It comprises three parts. The first part, the provider part, specifies *who* offers the service. The second part, the function part, specifies *what* the services does. In the third part, the accessibility part, we specify *how* the service can be accessed.

The service provider is the person overall responsible for the service. He may delegate some responsibilities to other persons, resulting in for example a technical owner who can help the consumer if he has technical problems, a functional owner who determines the functionality of a services based on his own ideas and requests from (potential) service consumer, and a commercial owner who determines the price for using the service.

The function part specifies what the service does. We specify the service type for making it possible to search for certain types of services, e.g. 'calculation services' or 'read services'. Additionally we are interested in what transaction(s) a service supports. Besides the input and output parameters we also specify errors, preconditions and postconditions. It is important to explicitly specify the semantics of the terms used to prevent semantic conflicts between provider and consumer. We also see the QoS constraints as a part of the function specification as it is also externally visible behavior.

The service usage part tells the service consumer how to access the service. We need information on where the service is located, usually the location has the form of a URL. Also, we need to know what protocols we need to use for communicating with the service. If the location is a URL of course we need the HTTP protocol for accessing it. Sometimes the HTTP protocol is enough, but more commonly we need more protocols and sometimes protocols are used that are not based on the WWW-standards. As getting the service "right" the first time is very hard, a service normally has multiple versions. Based on new insights

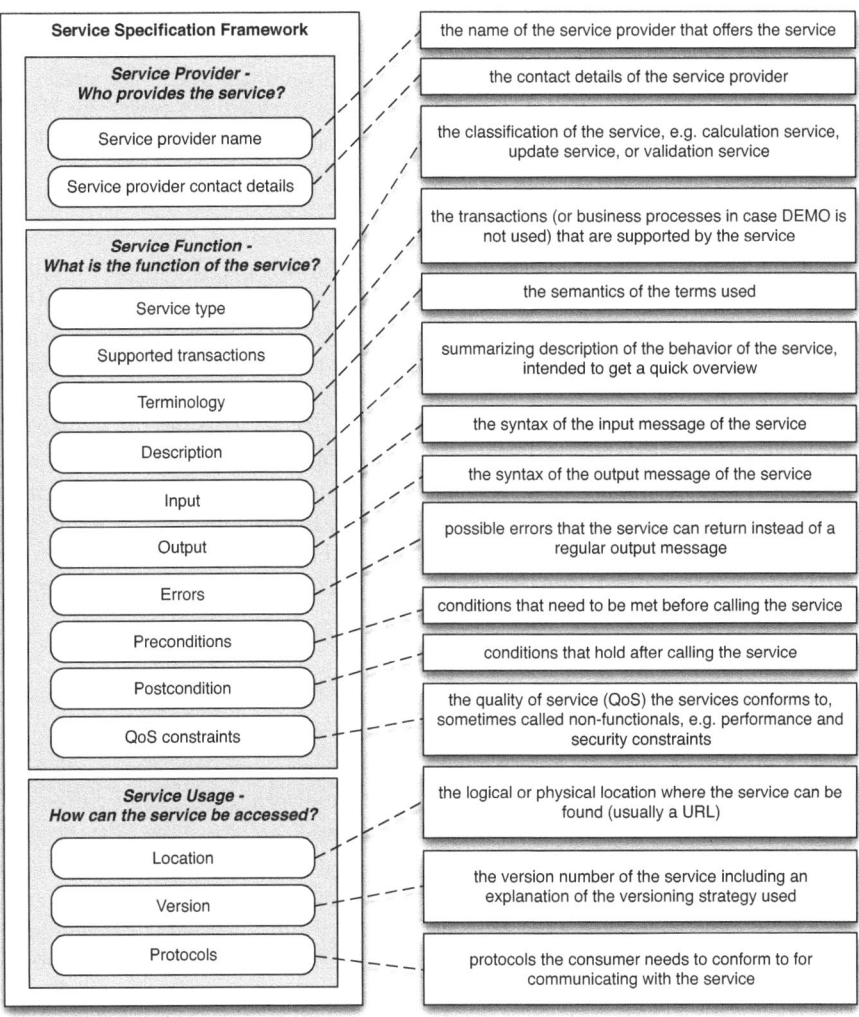

Fig. 3. Service Specification Framework

of the service provider and requests of service consumers the service provider may add, change or remove certain functionality. Adding functionality does not lead to compatibility issues as long as it is optional. Changing or removing certain functionality leads to incompatibility between service versions.

4   Case Study Background

The Port of Rotterdam Authority manages and develops the Rotterdam Port and Industrial Areas. It aims to: (i) promote effective, safe and efficient vessel traffic management, both within the port itself as well as in the coastal area and (ii) develop, construct, maintain and exploit port area's. Vessel traffic management is carried out by the *Harbor Master* Division, currently consisting of about 560 people, led by the Harbor Master. To perform its task, the Harbor Master Division has at its disposal: a central traffic management location, 3 (in near future 2) traffic control locations and 10 patrol vessels. Furthermore, a shore based radar system supplies a complete view of all vessels present in the port area. An extensive portfolio of information systems support the Harbor Master Division in many of its tasks. Due to the advanced age of the current main 'backbone' information system of the Harbor Master Division (resulting in costly maintenance and in long lead times for processing change requests), the decision has been taken to develop a new IT-system, named 'HaMIS' - Harbor Master Information System).

Our evaluation took place in the *Rational Unified Process* (RUP) elaboration phase. The elaboration phase has as goal to mitigate the main risks by testing the global design decisions, modeling the most important use cases, and building functionality of some of the use cases having the highest priority or being technically complicated. Another thing to do in the elaboration phase is setting up the development and test environment and preparing the customer organization for testing, accepting, and using the software.

5   Case Study Results

We have gathered data for five months through *participation-observation*, i.e. we have specified services ourselves using the framework, *direct observations*, i.e. we have looked what 'happened' and *interviewing*, i.e. we have evaluated the structure and contents of service specifications with a domain expert, a business analyst, four IT architects, and four software developers. In the evaluation we were looking for the answer to the following question: "To what extent do the aspects of the service specification cover the information needs of the project members?". For each aspect we want to indicate whether it was considered not applicable to the project, whether it was useful in an adapted form, or whether it was useful as is. Also, we want to find if the framework is missing any aspects required for service specification. In this section we first, in subsection 5.1, give an overview of the situation by presenting some of the identified services and by discussing parts of the specification of one service. After that, we discuss the evaluation results for each part of the framework in subsections 5.2 to 5.4. To give an impression of some of the situations encountered, we add several vignettes [19], i.e. briefly described episodes to illustrate an aspect of the case. The names used in these vignettes are fictional.

5.1   Identified Services

Table 1 exhibits a selection of the services identified during the first iteration of the RUP elaboration phase.

Table 1. Service identified during first iteration

Service name	Service description
SearchShipVisits	Search for ship visits using several input parameters
SearchBerthVisits	Search for berth visits using several input parameters
SearchShipMovements	Search for ship movements using several input parameters
GetShipVisit	Read the complete ship visit tree
SearchInspectionTasks	Search for inspection tasks, either assigned to the current user, available tasks, or, assigned to other users
ClaimInspectionTask	Assign an inspection task to current user or other user (depending on authorization level)
CompleteShipVisit-InspectionTask	Register the results of a ship visit inspection task
ReleaseInspectionTask	Undo the assignment of a certain inspection task
...	...

Let us have a closer look at the SearchShipVisits service as an example. Its description is: 'This service returns a list of ship visits conforming to given search criteria. It returns all attributes of the ship visit, all attributes of the ship belonging to the ship visit, a selection of the attributes of the agent representing the ship visit, and zero or one ship movement ids'. Table 2 shows the explanation of three of the terms from the terminology aspect in the service function part.

Table 2. The terminology aspect of the service function part

Ship visit	a non-interrupted stay of a ship in the certain geographical area defined by the harbor master.
Ship	a water vessel, including water airplane, hydrofoil, hovercraft, rig, production platform, dredger, floating crane, elevator, pantoon, and every floating tool, object and installation.
Port	An area for receiving ships and transferring cargo. It is usually found at the edge of an ocean, sea, river, or lake.
...	...

One of the input fields of this search service is the field StartDateTime. From the terminology aspect we know that this field is a datetime indicating the start of a ship visit. However, we need the additional information to know how to use this field in context of the search service. That is why we have added the following information to this input field: 'the StartDateTime and EndDateTime together specify a period within which part of the ship visit should exist. This means that a ship visit is returned if (NOT (ShipVisitEndDateTime<StartDateTime OR ShipVisitStartDateTime>EndDateTime)). By default the StartDateTime is the current date and time and the EndDateTime is undefined'.

In table 3 we can see two of the errors that can occur. Table 4 shows some of the QoS constraints to which the service conforms.

Table 3. The errors aspect of the service function part

Error code	Error name	Cause	Error message
F00000001	NoSearchElement	None of the input search elements are filled [unfulfilled condition on input].	"The service requires at least one search element as input"
F00000002	CharNotAllowed	A string search element contains characters that are not allowed, e.g. a wildcard in the middle of the string or Russian characters	"The name input element contains a character that is not allowed"
...	...	...	...

Table 4. QoS constraints aspect of the service function part

Characteristic	Sub characteristic	Constraint
Efficiency	Time behavior	The maximum response time in 90% of the calls is 1,75 seconds.
Efficiency	Resource behavior	The service can be called 50 times a minute.
...	...	...

The SearchShipVisits service uses SOAP and HTTP (as transport layer) as protocols, because for searching for information we do not need guaranteed message delivery. For the service ClaimInspectionTask, for instance, we would need guaranteed message delivery and we would choose Java Message Service (JMS) as a transport layer. Table 5 exhibits the locations of the service. We explain why we need multiple locations in subsection 5.4.

Table 5. The location aspect of the service usage part

Environment	URL
Development	123.456.789.123:5050/ShipVisit-v1/Service?wsdl
Test	123.456.789.456:5050/ShipVisit-v1/Service?wsdl
Acceptance	123.456.789.789:80/ShipVisit-v1/Service?wsdl
Production	123.456.789.321:80/ShipVisit-v1/Service?wsdl

5.2 Service Provider Part Evaluation

The service provider part (see Table 6) was not particularly useful in the project, because only one service provider existed. It would be nonsense to specify the

Table 6. Service Provider Part Evaluation

Aspect	Not applicable	Adapted	Useful
Service provider name	x		
During the case study period, it was unnecessary to specify the service provider information explicitly because the service provider was only one person.			
Service provider contact details	x		
The same finding holds for the service provider contact details as for the service provider name.			

name of this person in all service specifications. The architects argued that service provider information will be required in a later phase of the project when the number of services grows. When this occurs, they would prefer to specify a certain role or department instead of a person, since a person can work part time, become sick etc.

5.3   Service Function Part Evaluation

The project members all agreed that the service function part covered all aspects of the externally visible behavior of the service that are relevant for this project. Table 7 exhibits the results per aspect of the service function part.

Because we did not use the notion of Enterprise Ontology (with its accompanying method DEMO) in this project for business modeling, we needed to adapt the 'supported transactions' part. In this project traceability was realized by specifying a relation between a service and a use case. This use case in its turn is related to a business process. Figures 4, 5, and 6 present vignettes related to resp. terminology, input & output, and preconditions & postconditions.

Table 7. Service Function Part Evaluation

Aspect	Not applicable	Adapted	Useful
Service type			x
Though we have specified the service types, e.g. search service, read service, and task service, it was not (yet) required in this project because the number of services is still quite limited. Searching for potential reuse will be important when the number of services grows.			
Supported transactions		x	
As the HaMIS project is a very large software development project, traceability of why services are needed is required. In the project services are related to use cases instead of transactions.			
Terminology			x
To prevent inconsistency in the message usage for interaction with different services, we have designed a canonical data model that specifies all possible data elements used in the interaction with the services. The input and output parameters of a services refer to data elements in this canonical data model. We guarantee compliance between the messages and the canonical data model by only allowing the input and output messages being specified in terms of *restrictions* on the canonical data model.			

Table 7. (*continued*)

Aspect	Not applicable	Adapted	Useful
Description			x
This aspect is used for giving a summary of what the service does. Its use is to get a quick picture of the behavior of a service without having to read all the details.			
Input			x
We tried two ways of specifying the input parameter, viz. specifying them using UML class diagrams and in graphical representations of the XML schema trees. It turns out that architects prefer the first way of representation because they are only interested in what information is exchanged and not in the precise structure of the XML messages. Developers prefer the schema trees, or, the textual representation of the XML schemas. We used additional descriptions in natural language to specify conditions on the input parameters, e.g. if parameter a is empty, then parameter b must be filled.			
Output			x
The same findings holds for the output as for the input.			
Errors			x
Specifying the error situations was regarded as one of the most important aspects of the service specification. The provider and consumer need to create an understanding of what can go wrong. When getting a certain error message the consumer needs to know whether it is useful to call the service again with the same input parameters, whether he need to change his input parameters, or whether there is nothing he can do about it (in that case he needs to contact the service provider).			
Preconditions			x
We specified the preconditions in natural language. A discussion arose on how to deal with pre-conditions and postconditions (see Fig. 6)			
Postconditions			x
The same findings holds for the postconditions as for the preconditions.			
QoS constraints			x
QoS constraint are an essential part of the service specification in the HaMIS project. We have used the Extended ISO model [22], an extension to the ISO 9126 model, as a basis for specifying the QoS constraints. Not all elements of this model are relevant, e.g. the usability element only applies to user interfaces and not to services. For this reason we have made a selection of elements of this model that need to be specified. Time behavior, availability, and security elements were regarded as the most important QoS constraints.			

Service specifier Iris asks domain expert Susan and business analyst Charles to provide a definition of the term 'tonnage' (of a ship). Its value is not always expressed in thousand kilograms, like the name seems to imply. Sometimes the value is expressed in units of 1016 kg, 1.1 cubic meter, or 2.8 cubic meters depending on which type of ship it regards and which maritime organization delivers the value. Susan proposed to "just use an integer", because the end user has enough domain knowledge what the value means in a specific context. It would not lead to problems because the only consuming component is the GUI. However, it can lead to serious errors when another consuming components starts making calculations with the tonnage value having its own assumptions on its semantics.

Fig. 4. Vignette 'The semantics of a Ship's Tonnage'

Architects Ellen and John require a service 'Search for ship visits'. This service returns ships that visit the port at a certain time (past, present, or future). Domain expert Susan and business analyst Charles bring up the required search criteria. Though this seems very straightforward, the XML Schema and WSDL standards are insufficient for specifying the input and output parameters. For instance, Susan wants to allow wildcards in search queries, e.g. 'Ship = HMS Beag*'. Also, she wants some input to be conditional, e.g. if input item 'Berth' is filled than also input item 'StartDateTime' and 'EndDateTime' should be filled. Because XML Schema lack the means to specify these details, we need some additional input and output descriptions in these service catalog.

Fig. 5. Vignette 'XML Schema Only is Insufficient for Specifying Input and Output'

In the service catalog Iris specifies the pre-and postconditions in natural language, since the software engineers are not familiar with more formal approaches like UML OCL, Z, or SWRL. Despite the lack of precision of expressions in natural language, both Rick (software engineer of the providing component) and Chris (software engineer of the consuming component) have the same understanding of the semantics of the expressions themselves. However, architects Dave (of providing component) and John (of consuming component) started a discussion about how to deal with these pre-and postconditions. Dave argued that in a Design by Contract^{TM} the caller is completely responsible for checking the preconditions. Not fulfilling them leads to undefined behavior (the service may or may not carry out its intended work). John voted against taking this approach. He opted for a double check at the service provider side, making sure a unfulfilled precondition always results in an error message returned by the service provider. His motivation was that an undefined output can jeopardize the functioning of the complete HaMIS system. Though he sees the double work for implementing condition checking at both sides, he sees undefined output as an unacceptable risk.

Fig. 6. Vignette 'How to Deal with Pre- and Postconditions'

5.4   Service Usage Part Evaluation

Table 8 shows the results per aspect of the service usage part. From the interviews we found that the aspects of the service usage part sufficiently addressed the information needs of the architects and software developers. The other project members did not need the service usage part as they were only interested in the function of the service. The architects and software developers proposed a change to the service location aspect. In our original service specification template 'location' referred to just one location, in the project we needed to specify multiple locations. In software development projects in general a clear distinction is made between the following type of physically separated environments: Development, Test, Acceptance, and Production. They each have their own purpose within the software development process. Something new or a change to existing software should be developed in the Development environment. It should be tested in the Test environment by the project members. After that it should be tested by a selected group of end users in the Acceptance environment. If this group of end

Table 8. Service Usage Part Evaluation

Aspect	Not applicable	Adapted	Useful
Location		x	
In this project specifying only one location for a service was deemed insufficient. Instead, we needed a Development, Test, Acceptance, and Production location.			
Version			x
Because during a software development project services change quite often, a good service versioning mechanism is crucial. For this project we have applied the backwards compatibility strategy as defined by Erl et al. [7] to the canonical data model as well as the messages. This results into the following type of version numbers: "x.y", in which x represents a major version number and y a minor version number. The terms major and minor relate to compatibility with previous versions, for instance: version 5.3 is compatible with version 5.1, but not with version 4.8. This was considered to be sufficient.			
Protocols			x
For accessing the service we use the WSDL standard. A web service can be built using RPC or document style binding. A few members of the project team, viz. two architects, one developer, two integration specialists, and the service specifier choose to apply document literal style web services for their message validation capabilities, the possibility to define XML schemas externally (outside the service interface description) and their WS-I compliance (conformation to standards) [2]. Though a drawback may be a lesser performance, we think the benefits outweigh this drawback. Because all services will use a document style binding we do not need to make this information explicit in the individual service specifications. In our project we use two types of transport layers: HTTP (for synchronous service calls) and JMS (for asynchronous service calls). Though the difference can be seen in the binding part of the WSDL, we also specify this in the service specification for making this information available to other stakeholders than developers.			

users accepts the software, it can be propagated to the Production environment in which the software is used by the actual end users. When tests fail in either the Test or Acceptance environment, the software is demoted back to the Development environment and the process restarts. Since the services play a role in all these environment, their locations for all these environments need to be specified.

5.5   Overall Evaluation

All in all, the architects and software engineers from the providing as well as the consuming party expressed their enthusiasm about the service specifications. The software engineers saved time because both parties agreed upon the external behavior in an early stage. This enabled both parties to work in parallel; the consumer used stubs of the actual service. Replacing the stub by the actual service led to no or very few problems. When problems did occur, it was always immediately clear by looking at the service specification which party has caused the problem(s).

According to the interviewees not only the framework itself contributed to the prevention of errors and early error discovery, but also the structured specification process and the separate role of 'neutral' service specifier. By neutral we mean that the service specifier does not work on the design or implementation of either the providing or the consuming components. Because of this we prevent component-specific constructs in the specification ('shortcuts') for making the implementation easier.

6 Conclusions

In this paper, we sought to evaluate our service specification framework in a real-life case study at the Port of Rotterdam. This case study contributed to our ultimate goal, i.e. creating a generic service specification framework that is both founded on a sound scientific theory (the Ψ theory) and evaluated in several real-life projects. We position this case study as the evaluation step of the design science research methodology of Peffers et al.

The main contributions of this framework in the HaMIS project consisted of early error discovery and awareness of important service aspects that are often overlooked. These errors mainly included semantic errors and conditions to the input messages that cannot be specified using XML schemas and WSDL. The QoS constraints aspect in the service specification framework made people aware that this is also externally visible behavior of a service. It lead to negotiations about for instance response time, availability, and security between service provider and consumer. Also, people became aware that it is not only important to specify the 'happy scenario', but to also take into account the specification of different types of error situations.

Overall the service specification framework fulfilled the needs of the HaMIS project, though some findings led to improvements in our framework. These findings include the following: (i) the need for specification of multiple service locations, (ii) the need for specification of roles or departments instead of people in the service provider part, (iii) the need for specification of conditional input and output parameters, (iv) the need for making explicit how to deal with unfulfilled preconditions.

References

1. Ackermann, J., et al.: Standardized specification of business components (February 2002),
   http://www.wi2.info/downloads/gi-files/MEMO/
   Memorandum-english-final-included.pdf
2. Akram, A., Allan, R., Meredith, D.: Best practices in web service style, data binding and validation for use in data-centric scientific applications. In: Proceedings of the UK e-Science All Hands Meeting, Nottingham, UK (September 2006) Imperial College of London
3. Albani, A., Dietz, J.L.G.: The benefit of enterprise ontology in identifying business components. In: Proceedings of WCC, Santiago de Chile, Chile (2006)
4. Booch, G.: Object-Oriented Analysis and Design with Applications, 2nd edn., Benjamin–Cummings, Redwood City, California, USA (1994)
5. Clements, P., et al.: Documenting Software Architectures. Pearson, London (2002)

6. Dietz, J.L.G.: Enterprise Ontology, Theory and Methodology. Springer, Heidelberg (2006)
7. Erl, T., et al.: Web Service Contract Design and Versioning for Soa. Prentice Hall PTR, Upper Saddle River (2008)
8. Flores, F., Ludlow, J.: Doing and speaking in the office. Decision Support Systems, Issues and Challenges, 95–118 (1980)
9. Goldkuhl, G., Lyytinen, K.: A language action view of information systems. In: Proceedings of the ICIS, Ann Arbor, MI, USA, pp. 13–29 (1982)
10. Habermas, J.: Theorie des kommunikativen Handelns. Suhrkamp Verlag, Frankfurt am Main (1981)
11. Hevner, A.R., March, S.T.: The information systems research cycle. Computer 36(11), 111–113 (2003)
12. Hevner, A.R., March, S.T., Park, J., Ram, S.: Design science in information systems research. MIS Quarterly 28(1), 75–105 (2004)
13. Jacobson, I.: Object-Oriented Software Engineering: a Use Case driven Approach. Addison-Wesley, Wokingham (1995)
14. Levi, K., Arsanjani, A.: A goal-driven approach to enterprise component identification and specification. Communications of the ACM 45(10), 45–52 (2002)
15. McGovern, J., Sims, O., Jain, A., Little, M.: Enterprise Service Oriented Architectures. Springer, New York (2006)
16. Parnas, D.L.: On the criteria to be used in decomposing systems into modules. Communications of the ACM 15(12), 1053–1058 (1972)
17. Peffers, K., et al.: A design science research methodology for information systems research. Journal of Management Information Systems 24(3), 45–77 (2007)
18. Rumbaugh, J., et al.: Object-oriented modeling and design. Prentice-Hall, Inc., Upper Saddle River (1991)
19. Stake, R.E.: The art of case study research. Sage Publications, Thousand Oaks (1995)
20. van Aken, J.E.: Management Research Based on the Paradigm of the Design Sciences: The Quest for Field-Tested and Grounded Technological Rules. Journal of Management Studies 41(2), 219–246 (2004)
21. van Strien, P.J.: Towards a methodology of psychological practice: The regulative cycle. Theory Psychology 7(5), 683–700 (1997)
22. van Zeist, B., Hendriks, P.: Specifying software quality with the extended ISO model. Software Quality Journal 5(4), 273–284 (1996)
23. Yin, R.K.: Case Study Research: Design and Methods, 3rd edn. SAGE Publications, Thousand Oaks (2002)
24. Zhang, Z., Liu, R., Yang, H.: Service identification and packaging in service oriented reengineering. In: Proceedings of the SEKE, pp. 241–249 (2005)

Enhancing the Formal Foundations of BPMN by Enterprise Ontology

Dieter Van Nuffel[1], Hans Mulder[1], and Steven Van Kervel[2]

[1] Department of Management Information Systems, Universiteit Antwerpen,
Prinsstraat 13, B-2000 Antwerpen, Belgium
dieter.vannuffel, hans.mulder@ua.ac.be
[2] Department of Information Systems, Delft University of Technology, Mekelweg 4,
2628 CD Delft, The Netherlands
steefk22@telenet.be

Abstract. Recently, business processes are receiving more attention as process-centric representations of an enterprise. This paper focuses on the Business Process Modeling Notation (BPMN), that is becoming an industry standard. However, BPMN has some drawbacks such as the lack of formal semantics, limited potential for verification, and ambiguous description of the constructs. Also the ontology used to model is mostly kept implicit. As a result, BPMN models may be ambiguous, inconsistent or incomplete. In order to overcome these limitations, a contribution to BPMN is proposed by applying the way of thinking of DEMO; the explicit specified Enterprise ontology axioms and the rigid modeling methodology of DEMO. Adding the ontological concepts which, in DEMO, are translated into a coherent set of modeling symbols, may result in formal, unambiguous BPMN business process models. As such BPMN can be enriched on several aspects like the diagnosis, consistency, and optimalization of business processes.

Keywords: Enterprise Engineering, Enterprise Ontology, DEMO, Business Process Modeling, BPMN.

1 Introduction

In the last decade, business processes are receiving more attention as process-centric representations of an enterprise. Whereas earlier, mostly data-driven approaches have been pursued as starting point for information systems modeling, there is currently a tendency to use more process-driven requirements engineering [14]. This trend has even increased by recent developments like Service-Oriented Architectures (SOA) where business process languages are considered as primary requirements sources [16]. To model business processes, a large number of notations, languages and tools exist. This research focuses on the Business Process Modeling Notation (BPMN) [8] because BPMN is used to capture business processes in the early phases of systems development [4]. BPMN is moreover a quite intuitive notation that can become an official process modeling industry standard [11].

A. Albani, J. Barjis, and J.L.G. Dietz (Eds.): CIAO!/EOMAS 2009, LNBIP 34, pp. 115–129, 2009.

However, business process languages in general, and BPMN in particular, have some drawbacks. Regarding the basic assumptions of current business process modeling languages, following remarks are mentioned: absence of formal semantics, limited potential for verification, message-oriented approach, and the modeling of multi-party collaborations [2]. When analyzing BPMN, the lack of formal semantics is caused by the heterogeneity of its constructs, and the absence of an unambiguous definition of the notation [4]. In contrast with the comprehensively documented syntactic rules, the semantic meaning of the constructs is dispersed troughout the specification document in plain text [4]. Although BPMN is a relatively recent standard, it has already been evaluated using a number of theories and frameworks[1]. The following overview of BPMN's evaluations mainly concentrates on the aspects relevant for our research: the facts concerning completeness, consistency and ambiguity of the modeled business processes using BPMN. BPMN may have additional strenghts and weaknesses, but these are not discussed hereafter.

First, BPMN was analyzed using the Workflow Patterns [15]. The results indicated that the data perspective is not fully covered as opposed to the control flow perspective which is quite extensively supported. A second evaluation was made by means of the Representation Theory using the Bunge-Wand-Weber (BWW) ontology as a framework [9,10]. Following findings were reported:

- Concerning ontological completeness, it is concluded that BPMN lacks representations of state and history.
- Regarding construct excess (i.e. BPMN constructs not representing any BWW ontological artifact), a number of BPMN artifacts have no real-world meaning, for instance the Text Annotation construct.
- Concerning construct overload (i.e. more than one BWW ontological artifact maps to a BPMN artifact), lanes and pools map to multiple BWW constructs.
- Regarding construct redundancy (i.e. one BWW ontological artifact is represented by more than one BPMN construct), a thing can be represented by both a pool and a lane. A transformation is represented by an activity, a task, a collapsed sub-process, an expanded sub-process, and a transaction. An BWW event can be represented in BPMN by a start event, intermediate event, end event, message event, timer event, error event, cancel event, compensation event, and terminate event.

In comparison with other modeling techniques evaluated by the BWW ontology, BPMN appeared to be highly ontological complete, but inferior regarding construct clarity. Therefore the use of BPMN will lead to quite complete, but unclear and potentially ambiguous representations of real-world domains. In addition, the two above mentioned evaluation frameworks seem to be complementary as it is suggested that the workflow patterns are suited for evaluating the workflow view, whereas Representation Theory is useful to check the individual constructs [11].

---

[1] It should however be mentioned that these evaluations have been done on BPMN version 1.0.

A third evaluation framework, the Semiotic Quality Framework [6] that is based upon seven general quality aspects, identifies five criteria to assess the quality of conceptual modeling languages. Applying this framework to BPMN suggests that BPMN can easily be learned for simple use, and is easy to understand [13]:

- Domain Appropriateness (how suitable is a language for use within different domains): BPMN is suited to model the functional perspective. However, it is not suited to model functional breakdowns, business rules, and data and information models.
- Participant Language Knowledge Appropriateness (participants know the language and are able to use it): graphical elements of BPMN are clearly defined and easy to learn.
- Knowledge Externalizability Appropriateness (participants' ability to express their relevant knowledge using the modeling language): BPMN is appropriate for business to model business processes.
- Comprehensibility Appropriateness (audience should be able to understand as much as possible of the language): this category can be divided into understanding the language concepts, and understanding the notation. Regarding the latter, readers can easily recognize the basic types of elements as these types are limited in number, intuitive, and very distinguishable from each other. Regarding the language concepts, it is suggested that these are descriptive, accurate, easy to understand, and well defined.
- Technical Actor Interpretation Appropriateness (language suitable for automatic reasoning): it is said that business process diagrams (BPD) are "with a few exceptions easily translated into BPEL" [13].

Concluding the BPMN evaluations, it can be mentioned that the theoretical studies indicates more problems than practitioners actually suffer from [10]. Most problems were moreover indicated by people with an IT background who need more rigor and details to use BPMN models as input for software implementation projects; whereas business people pointed out that the use of the core set is sufficient and convenient for modeling concise models, easy to understand by business. This is also illustrated by the fact that only a subset of BPMN constructs is actually used when building business process diagrams [17].

Given the fact that BPMN is becoming the industry standard for business process modeling, but has some significant drawbacks mainly regarding ambiguous and unclear descriptions of their constructs, the contribution of this paper is to provide a formal foundation based upon which BPMN models with less ambiguity can be created. The focus of the paper is thus put on the business-oriented use of BPMN to model the business processes of an organization, as the proposed solution will be targeted at the early mentioned aspects of consistency, completeness and ambiguity.

The remainder of the paper is structured as follows. In the second section, BPMN is applied to an example and lessons learned will be discussed. The third section will give an overview of the Enterprise Ontology and DEMO methodology providing the basis for a rigid methodology to construct BPMN models with less or no ambiguity. Also practical recommendations will be given. Finally, conclusions and future research will be presented in the fourth section.

2   Case Study

The approach put forward in this paper will be illustrated by means of an example: the first phase of a pizzeria called Mama Mia [3, p.221-223]. Hereafter the relevant part of the case is described with some text formatted: either bold, italic or underlined. The aim of the formatting will be explained in the next section.

> "Customers **address themselves** to the counter of the pizzeria or make a telephone call. In both cases Mia <u>writes down</u> *the name of the customer, the ordered items, and the total price* <u>on an order form</u>. On the counter lies a <u>plasticized list</u> of the *available pizza's and their prices*. Usually she **produces** this list every year during their holiday. In case of an order by telephone she also <u>records</u> the *telephone number*. Moreover, she *repeats the ordered items* and *informs the customer about the price and the expected time* that the order will be ready. If necessary, she also *tells the customer the assortment of pizzas*. The order forms have a serial number and are produced in duplicate: a white and a pink copy. Mia <u>shifts the pink one</u> through a hatch in the wall to the kitchen, where Mario takes care of **baking the pizzas**. She keeps the white copy behind the counter. As soon as Mario **has finished an order**, he <u>shifts</u> the pizzas in boxes through the same hatch to Mia, including <u>the pink order copy</u>. Mia then seeks the matching white copy, <u>hands it</u> together with the boxes <u>over to the customer</u>, and waits for **payment**. It may happen that Mario is not able to fulfill an order completely because of missing ingredients. In such a case he puts his head through the hatch and *notifies Mia of the problem*. He then also <u>returns the pink copy</u>. If the customer is present in the shop, **she confers with him or her** what to do about it, and modifies the order. If the customer is not present, which is mostly the case for telephonic orders, she **modifies the order to her own discretion**. This leads sometimes to vigorous debates in the pizzeria when the customer comes for taking away the order. Thanks to Mia's temperament she always **comes to an agreement** that is not disadvantageous for her."

In figure 1 the business process of the pizzeria is modeled in BPMN. It should be mentioned that abstraction is made from the customer choosing a pizza. Modeling this choice communication would unnecessarily complicate the model, as it is not needed to illustrate the paper's contribution. When modeling the business process, several issues occured, mainly due to the ambiguous semantic meaning of the BPMN constructs. For example, how should you optimally model the payment request from Mia to the customer? Two options exist: first, you can opt to model it as an intermediate message event; second, it is possible to model it as an activity. The same ambiguity was also present when modeling the "Process Pizza" subprocess: which activities should be part of this subprocess? These issues are indicators of the construct redundancy present in BPMN. More generally, which activities should be part of the business process model

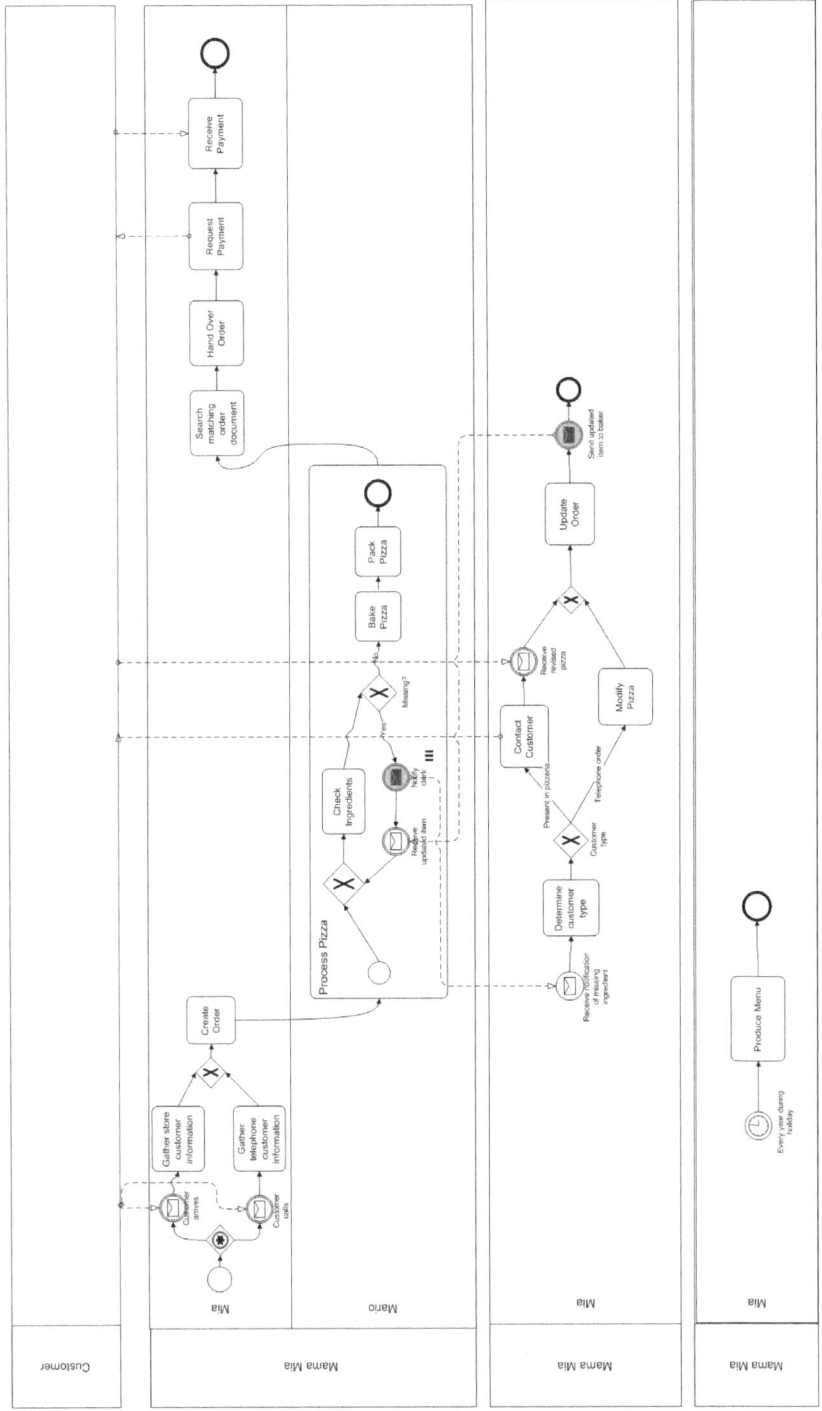

Fig. 1. BPMN diagram of the Pizzeria example

is a subjective opinion of the individual modelers. Aside from revision by end users or other stakeholders, there is no foundation to check whether the model is complete. Furthermore there are multiple options to model contacting the customer when some ingredients are missing. Because Mia is the intermediary that has to transfer the customer's preference to Mario, this interaction should be modeled within Mia's lane. Besides the option used in figure 1, one could also use signalling intermediate events. Again, no preference could be identified in the specification. Finally, the lack of support for the data perspective prohibits a clear overview of the process. For example, the only option that is available to state that an order consists of one or more pizza's, is using a text annotation. It should however be mentioned that the multiple instance indicator of the "Process Pizza" subprocess suggests this multiplicity in some way.

3   Solution Approach

To add a formal foundation to BPMN models, our approach is based on the Enterprise Ontology theory and DEMO, a methodology derived from Enterprise Ontology. In the first subsection, we will discuss the relationship between DEMO and BPMN in a general way, whereas in the second subsection DEMO will be applied to the BPMN model of the case study.

3.1   Enterprise Ontology and DEMO

DEMO is a methodology to construct enterprise models based on Enterprise Ontology (EO), an ontological theory [3]. The notion of ontology is defined as "an explicit specification of a conceptualization" [5]. EO has a strong ontological appropriateness because it has theoretical foundations in philosophy and sociology; is based on a set of three precisely formulated axioms and one theorem; and it is accepted by the community and large enterprises. DEMO models resulting from the EO-based DEMO methodology exhibit the following characteristics [3]:

- *Abstract*: DEMO models aim at a shared conceptualization for all stakeholders.
- *Formally correct*: Coherent, Comprehensive, Consistent, Concise and Essential ("C4E"). The formal correctness assures a guaranteed correct and identically shared abstract conceptualization for all stakeholders, while eliminating any ambiguity in interpretation. The formal correctness allows and is a mandatory condition for automated construction of information systems.
- *Essential*: DEMO incorporates a rigid and clear separation of an essential ontological model without any non-essential implementation oriented detailing.
- *Unique*: Model validation results in one and only one correct model for any enterprise.

Because BPMN models are composed of concepts, there must be some underlying domain ontology. However, BPMN does not prescribe ontologies for business, information and data processes, and their results. Moreover, designers of BPMN models are free to choose their ontologies; and, in practice, the applied ontologies

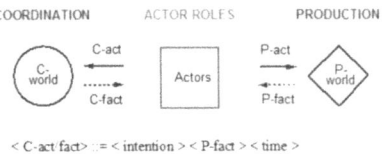

Fig. 2. Operation Axiom

are mostly kept implicit. Combining these two facts, it can be argued that Enterprise Ontology can be used as underlying domain ontology for BPMN models, exhibiting a high level of ontological appropriateness based on the arguments that the concepts of BPMN models are "similar" to the concepts of EO:

- The concept of an activity in BPMN is "similar" to the concepts of either a performance or DEMO Production (P-) Fact; or some communication about the performance or DEMO Communication (C-) fact.
- The concept of a lane in BPMN is "similar" to the concept of an Actor (Actor-Initiator or Actor-Executor).

To ground this hypothesis, both an evaluation of the EO ontological concepts, and a mapping to the primitives or concepts of BPMN have to be made. We start by evaluating the ontological concepts of Enterprise Ontology. EO is based on three axioms: the Operation Axiom, the Transaction Axiom and the Composition Axiom. The EO axioms are specified in schemes, i.e. formal graphical languages, and in natural language. A careful analysis is needed to identify all elementary and true propositions in natural language. An obvious requirement for completeness is that each entity in the EO axioms is identified. The operation axiom, illustrated in figure 2 states that there are Actors, performing some production or performance, and the actual production requires a coordination between these Actors. There are relations between the elements of the set of Actors, the C-world containing coordination facts or C-facts, and the P-world containing P-facts [3]. Ontological appropriateness is supported by the observation that any performance or production involves an Actor(role), and some form of communication with another Actor(role). As such, five propositions can be derived:

- *Proposition 1*: There are identifiable Actors, fulfilling roles where an Actor refers to a role, and is not directly linked to an identifiable natural Actor.
- *Proposition 2*: There are identifiable Pfact(s), representing some specific performance.
- *Proposition 3*: There are Cfact(s), representing communication about a specific performance to be delivered.
- *Proposition 4*: Each Pfact has a relation to one and only one unique Cfact, and vice versa.
- *Proposition 5*: Every Actor role, except for the root Actor-initiator, has one and only one Actor-Executor relation to one and only one Pfact, while it may have multiple (0 .. n) Actor-Initiator relations to other (child) Pfacts.

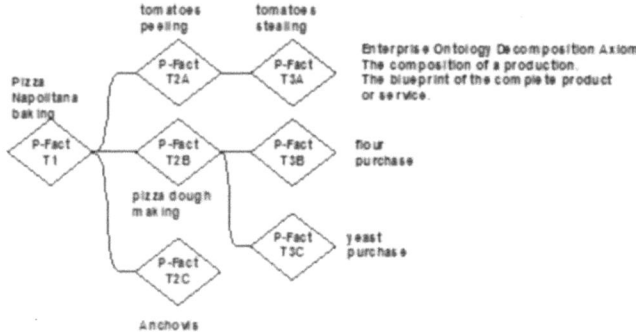

Fig. 3. Example of the Composition Axiom

The Composition Axiom, exemplified in figure 3, states that Pfacts can mostly be ordered in a hierarchical parent-child tree structure, representing the fact that for a certain Pfact to be performed, first a number of Child-Pfacts representing performance or production components, must have been performed in a recursive way. Ontological appropriateness is supported by the observation that many performances or productions require component performances or productions to become available first, as shown in figure 3.

The Transaction Axiom, shown in figure 4 in its most simple form of the basic transaction on the left and the standard transaction on the right, states that a transaction involves two Actor (roles), and follows a precisely specified pattern with states and state changes. The ontological appropriateness of the transaction axiom is strong: nested transactions up to unlimited nesting levels are computable and hence roll-back compatible. The following predicates can thus be added to the earlier mentioned propositions:

- *Proposition 6*: There is a constructional decomposition type of relation between a specific Pfact and any number (0 .. n) of child Pfacts.
- *Proposition 7*: For tangible Pfacts, all the child Pfact(s) have to be performed (Stated and Accepted) before the performance of the parent Pfact can start (Execution Phase).
- *Proposition 8*: An Actor with a Actor-Executor relation to a specific Pfact has an Actor-Initiator relation with each child Pfact.
- *Proposition 9*: There is one and only one Pfact, the root Pfact, that has a relation to an Actor that has exclusively an Actor-Initiator relation to this root Pfact.
- *Proposition 10*: There is at least one Pfact in a model without child Pfacts, i.e. a terminal Pfact, that has a relation to an Actor that has exclusively an Actor-Executor relation to this terminal Pfact.
- *Proposition 11*: There is a set of eight attributes (Request, Promise, etc.) uniquely related to specific Cfact, that describes the current state of communication of that element about the performance of the related Pfact.

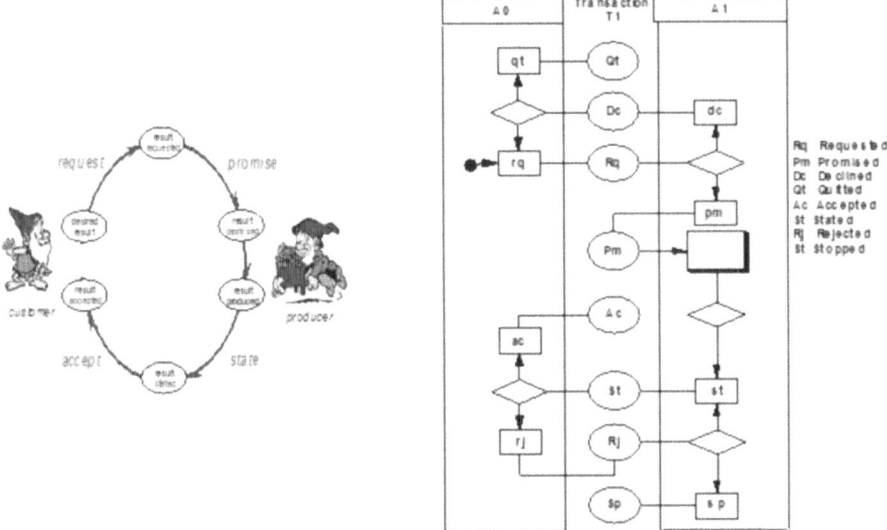

Fig. 4. Transaction Axiom, the basic and the standard transaction patterns

Investigating the relation between EO concepts and BPMN concepts yields the following conclusions with respect to BPMN. First, there are no transactions, transaction states and transaction state transitions specified in BPMN. Second, the relationship between an activity, the actor that requests the activity and the actor that performs the activity is optional in BPMN. Third, there is no separation between essence and implementation. As this latter needs more clarification, we first discuss the DEMO methodology in more detail.

Enterprise engineering starts by collecting information, all seemingly valuable statements about the purpose, construction and operation of the enterprise to be modeled. The second stage consists of applying the distinction theorem to separate and to discard any information that is not ontological, i.e. all data-logical and infological information is discarded. The remaining information is ontological, being the essence of an enterprise. Datalogical and infological information are considered to be implementation. Applying this step to the pizzeria case results in formatting the text: the bold text refers to the ontological information, the italic text to the infological information, and the underlined text to the datalogical information. The third stage investigates the remaining ontological information to identify ontological transactions and actors fulfilling either an Actor-initiator or Actor-executor role related to the respective ontological transaction. Enterprise engineering is about the question **what** has to be performed by the enterprise under investigation, and **how** the **what** is constructed in terms of aggregated components, typically described by a blueprint or an assembly manual [3]. The design of the enterprise, **who** does **what**, is defined by the allocation of the responsibility of the performance of the parts of manufacturing to actors. During operation or simulation, once the enterprise has

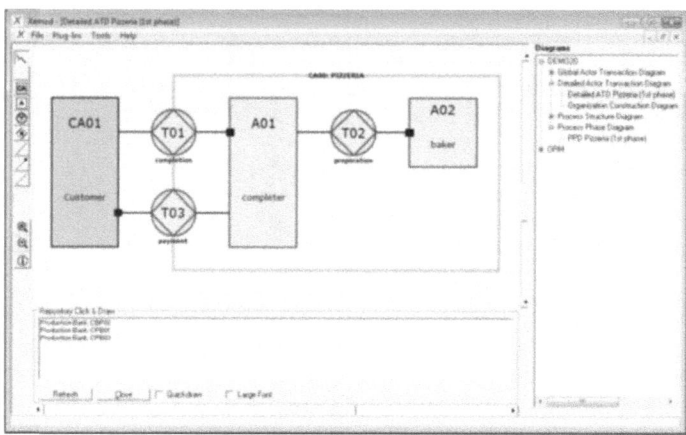

Fig. 5. Detailed ATD of the Case study

been constructed, the question what the current status of any production part is, must be resolved. Hence the specification of the DEMO model starts with the Pfacts, and their hierarchical ordening, representing the parts of the performance to be delivered. In this way the Actor-Transaction diagram (ATD), one of the 4 DEMO aspect models is constructed. In figure 5, the detailed ATD of the Pizzeria case study is shown [2].

The above-described methodology delivers "elementary" DEMO models: only transactions and actors are modeled, resulting in the elementary DEMO ATD. In further stages, several refinements and more detailed specifications are made, and other DEMO aspect models are specified. Relevant to our contribution is that the DEMO methodology is very rigid and concise, the steps and decisions to be taken to construct DEMO models are clearly specified, and do not allow ambiguity. There is only one correct validated DEMO model of an enterprise. As such, we argue that if there exists a domain ontology for a certain modeling notation, and if there exists a rigid methodology to derive models within that ontology domain, then that methodology yields models that should be represented by that modeling notation. Or rephrased: because EO is considered the domain ontology of BPMN, and DEMO is a rigid and concise modeling methodology to construct EO models, the DEMO methodology is also a rigid and concise modeling methodology for BPMN models. As EO and DEMO are formally correct, every BPMN model derived in this way should be formally correct. However, this line of reasoning has still to be validated by more research, and should therefore be considered as hypotheses guiding this work-in-progress. Besides questionning how BPMN models should be constructed based on the EO domain ontology and the DEMO methodology, there is also the question how existing BPMN models can be verified to be formally correct. BPMN is not a sentential language, but a

---

[2] The tool used to model the DEMO models is Xemod (Xprise Enterprise Engineering Modeler). More information: http://www.xemod.eu/

graphical language, so some methodology must be found to represent graphical BPMN models in a sentential form. If we have this sentential form, then formal correctness can be verified using a grammar. Such a grammar is currently researched by the authors.

To conclude, it can be stated that EO and DEMO have proven their strength for enterprise engineering, being the modeling of essential abstract enterprise models without any implementation details [7]. The main advantage is that a shared and formally correct conceptualization for all stakeholders has been established as the one and only right point of departure for any implementation. Preliminary results of research by the authors show that the EO and the DEMO methodology deliver formally correct models which allows the construction of a DEMO automata, the core of Enterprise Ontology derived information systems. Moreover, DEMO ends where BPMN starts. Therefore, a way to construct formally correct BPMN models by applying the DEMO methodology will be identified. As such, we gain the DEMO advantages and have a formal correct BPMN model. The formal correctness of the BPMN models can possibly allow automatic code generation in future. In addition, also methods to verify if a BPMN model meets the propositions of the EO axioms will be produced. In this sense, the research is related to the earlier work on business process analysis and simulation [1].

3.2   Practical Implications

In practice, there is a need for re-using existing diagrams within companies, thus in addition to the way of working in subsection 3.1, which can be characterized by a clean sheet approach, a reverse engineering of BPMN diagrams with DEMO is needed. A way of working to achieve this, is by manually classifying each of the BPMN elements within the business process models on the ontological, informational or documental level. The DEMO models focus on the Ontological level, BPMN also takes into account the informational and documental levels. The combination of both methodologies allows to represent, analyze and (re)design all levels, including the current organizational structures in relation to the ontological actors. This is illustrated by using the following part of the case: Mia writes down *the name of the customer, the ordered items, and the total price on an order form*. This consists of a document, namely the order form which can be modeled in BPMN, and a calculation of the total ordered items and price. Also a mapping of the current organizational structure in BPMN, to the ontological actor completer can be performed. In Figure 6, it is shown how to map the creation of the Order to the promise of transaction T01. When analyzing figure 1, it is noticed that the request is modeled through the message flow initiating at the border of the Customer pool. The reason for this is that the BPMN diagram is modeled from the perspective of the organization, thus receiving the Call or Arrival of the customer. In figure 7, the mapping of the Call or Arrival of the customer to the Request is performed.

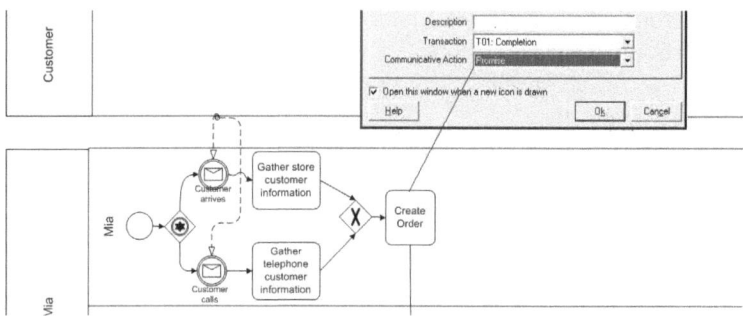

Fig. 6. Mapping of the Promise Order

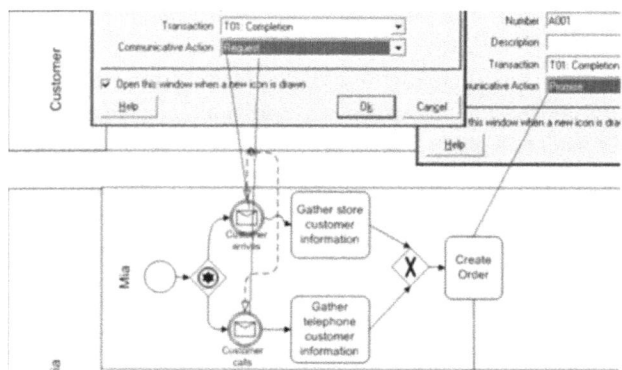

Fig. 7. Mapping of the Request Order

When performing this way of working, which is described in [3] as a performa-informa-forma analysis, a consistency check could be performed regarding the communicative actions. These consistency checks could involve:

1. Checking the completeness of the BMPN diagram on the ontological level (are all communicative actions mapped?)
2. Check if each communicative action has a documental and/or informational implementation (this means that the customer call is mapped on the ontological as well as the documental level)
3. Checking the implementation of Actors to persons/departments
4. Checking the sequence of the communicative actions (are these conform the DEMO process model?)
5. Check the implications for the BPMN diagram of a redesign of the DEMO process model

As researched for the Event-Driven-Process-Chain (EDPC) approach of ARIS and SAP [12], mapping of DEMO is valuable for enriching existing process descriptions. The approach is illustrated in figure 8, and has the following

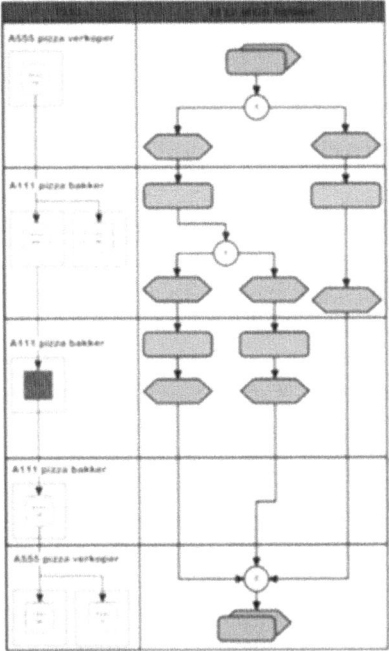

Fig. 8. Mapping of Actor to Persons/Departments [12]

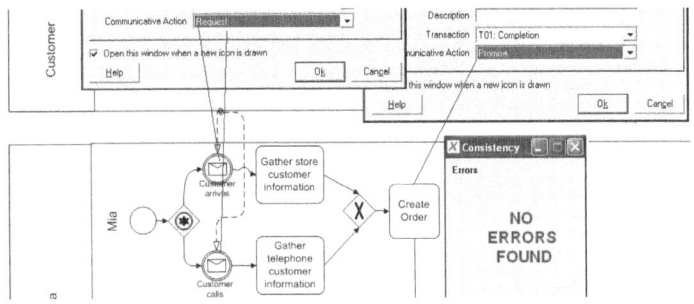

Fig. 9. Resulting illustration of approach

advantages: improved insight, improved visibility of business processes, improved checks on completeness, and on consistency. As already stated before, the same holds for BPMN. Applying the approach researched by Strijdhaftig, to the BPMN model of figure 1, leads to the results illustrated in figure 9. The presented approach in this paper thus wants to assure that business processes modeled within BPMN are systematically consistent with the enterprise modeled by the DEMO methodology. Strijdhaftig [12, p. 31-37] also mentioned the combination of DEMO with ARIS to provide a multitude of opportunities: "Basically, we know that the DEMO and ARIS architectures each describe aspects of the same organization and that there should be a systematic way to move from one architecture into the next and back again but the current way of modeling does not support that kind of transparency. [...] the benefit of allowing their users to move freely from one architecture to the other and see both views of the organization quickly and easily."

4   Conclusions and Future Research

This paper states that the design of enterprises, expressed by an ontology containing concepts of social interaction, in its implementation includes an ontology for rational information systems, and an ontology for documental systems. However most ontologies are implicit regarding the concepts constructing the realities of their to be represented systems, such as the enterprise and information systems. Therefore it can be argued from this preliminary research that documental and informational ontologies are subsets of an Enterprise ontology, leading to the conclusion that instead of investigating an one-to-one mapping, a many-to-many mapping has also to be further researched. The authors thus argue that Enterprise Ontology, combined with the DEMO methodology, can provide a formal foundation to BPMN models. Using EO as underlying domain ontology results in explicitly specified BPMN models, mainly because EO delivers constraints to which BPMN constructs should adhere. Moreover, revising existing BPMN models with DEMO can be used to verify completeness and consistency of the modeled business processes. The main contribution of the paper is thus combining the rich representational aspects of BPMN with the formal correctness of DEMO. These insights will be extended by providing defined ways to derive BPMN models from DEMO theory, and by producing tool support for the DEMO - BPMN integration.

References

1. Barjis, J.: Automatic business process analysis and simulation based on DEMO. Enterprise Information Systems 1(4), 365–381 (2007)
2. De Backer, M., Snoeck, M., Monsieur, G., Lemahieu, W., Dedene, G.: A scenario-based verification technique to assess the compatibility of collaborative business processes. Data and Knowledge Engineering (article in press) doi:10.1016/j.datak.2008.12.002
3. Dietz, J.L.G.: Enterprise Ontology: Theory and Methodology. Springer, Heidelberg (2006)
4. Dijkman, R., Dumas, M., Ouyang, C.: Semantics and analysis of business process models in BPMN. Information and Software Technology 50(12), 1281–1294 (2008)
5. Gruber, T.: Toward Principles for the Design of Ontologies Used for Knowledge Sharing. International Journal Human-Computer Studies 43(5-6), 907–928 (1995)
6. Lindland, O.I., Sindre, G., Solvberg, A.: Understanding Quality in Conceptual Modeling. IEEE Software 11(2), 42–49 (1994)
7. Mulder, J.B.F.: Rapid Enterprise Design. Ph.D. thesis, Delft University of Technology (2006)

8. Object Management Group: Business Process Modeling Notation, V1.2 OMG Available Specification (2009), http://www.omg.org/docs/formal/09-01-03.pdf
9. Recker, J., Indulska, M., Rosemann, M., Green, P.: Do Process Modelling Techniques Get Better? A Comparative Ontological Analysis of BPMN. In: Campbell, B., Underwood, J., Bunker, D. (eds.) Proceedings of the 16th Australasian Conference on Information Systems. Association for Information Systems (2005)
10. Recker, J., Indulska, M., Rosemann, M., Green, P.: How good is BPMN actually? Insights from practice and theory. In: Ljungberg, J., Andersson, M. (eds.) Proceedings 14th European Conference on Information Systems, pp. 1582–1593 (2006)
11. Recker, J., Wohed, P., Rosemann, M.: Representation Theory Versus Workflow Patterns - The Case of BPMN. In: Embley, D.W., Olivé, A., Ram, S. (eds.) ER 2006. LNCS, vol. 4215, pp. 68–83. Springer, Heidelberg (2006)
12. Strijdhaftig, D.: On the Coupling of Architectures: Leveraging DEMO Theory within the ARIS Framework. Master's thesis, Delft University of Technology (2008)
13. Wahl, T., Sindre, G.: An Analytical Evaluation of BPMN Using a Semiotic Quality Framework. In: Halpin, T., Siau, K., Krogstie, J. (eds.) Proceedings of the Workshop on Evaluating Modeling Methods for Systems Analysis and Design (EMMSAD 2005) (2005)
14. Weske, M.: Business Process Management: Concepts, Languages, Architectures. Springer, Heidelberg (2007)
15. Wohed, P., van der Aalst, W.M.P., Dumas, M., ter Hofstede, A.H.M., Russell, N.: On the Suitability of BPMN for Business Process Modelling. In: Dustdar, S., Fiadeiro, J.L., Sheth, A.P. (eds.) BPM 2006. LNCS, vol. 4102, pp. 161–176. Springer, Heidelberg (2006)
16. Zimmermann, O., Schlimm, N., Waller, G., Pestel, M.: Analysis and Design Techniques for Service-Oriented Development and Integration. In: Cremers, A.B., Manthey, R., Martini, P., Steinhage, V. (eds.) INFORMATIK 2005 - Informatik Live! Beiträge der 35. Jahrestagung der Gesellschaft für Informatik. Lecture Notes in Informatics, vol. 68, pp. 606–611. Springer, Heidelberg (2005)
17. zur Muehlen, M., Recker, J.: How Much Language is Enough? Theoretical and Practical Use of the Business Process Modeling Notation. In: Bellahsène, Z., Léonard, M. (eds.) CAiSE 2008. LNCS, vol. 5074, pp. 465–479. Springer, Heidelberg (2008)

Developing Quality Management Systems with DEMO

Jos Geskus[1,2] and Jan Dietz[1]

[1] Delft University of Technology
Faculty of EEMCS
Department of Software Technology
j.geskus@geskus.info, j.l.g.dietz@tudelft.nl
[2] INQA Quality Consultants B.V., The Netherlands

Abstract. The International Organization for Standardization (ISO) has defined Quality Management, but it has not yet adopted standards for developing Quality Management Systems (QMSs), notably not for modeling business processes in this context. Consequently a variety of modeling techniques are in use. Most of these are not able to produce concise and comprehensive models, whereas these features are particularly important for QMSs. Moreover, these techniques appear to be based on the mechanistic paradigm, meaning that they are task oriented instead of human oriented. Various researches indicate that this leads, among other things, to alienating employees from their work. DEMO (Design and Engineering Methodology for Organizations) has both desirable features: it is human oriented and it produces concise and comprehensive models of business processes, since it is based on the systemic notion of enterprise ontology. This paper reports on the theoretical evaluation of DEMO for the purpose of developing QMSs, as well as on practical experiences in applying DEMO to it.

Keywords: Quality Management System, DEMO, mechanistic paradigm, institutional paradigm.

1 Introduction

In general, the function of a company is to deliver particular services to its environment, where a service may regard tangible or intangible products. Also in general, companies strive to fulfill this function as good as possible, i.e. delivering high quality services. Quality, however, is a multi-faceted issue; it is not easy to measure it, nor to maintain or even improve it. Nonetheless, it is a major managerial area of concern in most companies. Collectively, this managerial attention together the activities performed to monitor, assess, and improve quality, are called Quality Management. In the realm of standardization, it is addressed by the ISO 9000 family of standards, in particular by the standard ISO 9001. In [1] Quality Management is defined as follows:

- fulfilling the customer's quality requirements, and
- the applicable regulatory requirements, while aiming to

A. Albani, J. Barjis, and J.L.G. Dietz (Eds.): CIAO!/EOMAS 2009, LNBIP 34, pp. 130–142, 2009.

- enhance customer satisfaction, and to
- achieve continual improvement of the organization's performance in pursuit of these objectives.

The focus in ISO's definition is on customer satisfaction. It is achieved by fulfilling the customer's quality requirements, which on its turn is enabled by business process improvement. The term 'regulatory requirements' remains rather unclear. Therefore we will not take it into account.

The implementation of ISO 9001 in a company can lead to an increase in productivity, customer satisfaction, less scrap and rework, and continuous development. Since most companies like to achieve these goals, ISO 9001 is applied in many of them [2]. Unfortunately, being compliant with this norm has become such a status symbol that quite some companies are proud to say that they have implemented the standard, whereas they have not really and fully done it [3]. It has been shown that most Total Quality Management (TQM) programs develop an "ideal organizational identity" that the enterprise presents to the outside-world, but which is often far removed from the daily reality [4].

DEMO (Design and Engineering Methodology for Organizations) is an enterprise engineering methodology. The authoritative source for DEMO is [5]. This paper reports on the theoretical evaluation of DEMO for the purpose of developing QMSs, as well as on practical experiences in applying DEMO to it. We have done that in the context of the new discipline of Enterprise Engineering. Although this discipline is certainly not fully established yet, the main characteristics are becoming clear [6]. They are summarized in the Enterprise Engineering Manifesto [7].

The remainder of the paper is structured as follows. Besides the problem statement are desired features for a QMS handbook given in Section 2. Section 3 contains an introduction to DEMO and explains how it fulfills the desired features stated in Section 2. Consequently the aspect models of the methodology that supports Enterprise Ontology are elaborated on. Section 4 explains the construction of a QMS handbook, the representations of the models are illustrated according a plain example. Finally evaluation and conclusions are to be found in Section 5.

2   Problem Statement

The objective, rational, and structured appearance that a QMS handbook usually shows and the concern for detailed activity descriptions that it addresses characterize the mechanistic and instrumental view of quality management on which the ISO 9000 standards are based. Moreover, the mechanistic viewpoint is shared by most of the studies of ISO 9000. The literature focuses primarily on describing the objective of this standard and its implications for quality management and organizations performace improvement [8,9,10,11]. In contrast to the mechanistic paradigm, the institutional paradigm attempts to describe how people interpret and construct their reality by conforming to standards and values that define "the way things are and/or the way things are to be done"

[12,13]. The latter seems to be interesting because the mechanistic view of organizations has been strongly contested by critical theory [14,15,16,17]. There are reasons to believe that the growing rationalization and formalization of human activities constitutes one of the dominant traits of modern society. The mechanistic view leads to the development of excessive bureaucratic organizations that are unwieldy and inflexible. Though the latter is considered to be precise and efficient, they are also characterized by impersonal and alienating employee control mechanisms [18,19,20,21,22,23]. Boiral's study shows that ISO 9000 is not a monolithic system that can be imposed in a mechanistic way. Rather, this standard is a socially constructed management system that people adopt, reinterpret, or reject depending on the situation. Boiral concludes furthermore that the importance given to the employees's participation and a common consensual view on the organization are key factors for a succeful implementation [3]. Practical experience of the authors confirms Boiral's conclusion. It turns out that commitment of the employees appears to be indispensable to a successful certification process. Commitment of the employees is also necessary guarding the quality management process [11,19]. Furthermore, Boiral [3] concludes that the focus in ISO 9000 process descriptions is on the production acts themselves and not on the interaction between employees. Focusing too much on the production acts (the mechanistic view) carries the risk of alienating employees from their work, which is of course an unwanted situation.

The predominance of the mechanistic paradigm in many organizations' statements and in studies of the ISO 9001 standard is subject of an ongoing dispute [3]. Therefore it might be interesting to apply a methodology to QMS that embodies the institutional paradigm. As said before, rationalizing and formalizing is necessary to achieve at precise and efficient process descriptions. Making processes transparant and involving employees in the design and implementation of the QMS will result in an increased commitment. It leads to less bureaucracy and it contributes to a successful implementation of the standard. Employee commitment during the design and implementation phase seems to a be very important factor. It determines whether a QMS implementation will be successful and whether the QMS is fully supported by the employees after the implementation, such that quality is guaranteed and improvement takes place. To receive the commitment of employees, the threshold to read the process descriptions in the QMS should be low. In order to lower this threshold the handbook must be concise. It must contain clear process descriptions, while irrelevant processes should be excluded. From the discussion above we arrive at the next problem statement:

> The introduction of Quality Management in an organization, in particular the design and implementation of a QMS, often evokes considerable resistance among the employees. There appear to be two main causes for this. One is that the selected approach is based on the mechanistic paradigm. The other cause is that the produced process models lack conciseness and comprehensiveness. It is hypothesized that applying an approach that is based on the institutional paradigm, as well as on the systemic notion of Enterprise Ontology, would take away these causes.

In order to achieve the expected benefits of applying the institutional paradigm, the following features are desirable for a QMS:

1. Conciseness of the handbook;
2. Describe the main line only;
3. No irrelevant processes;
4. No ambiguity in process description;
5. Not sensitive to minor process changes.

DEMO is currently the only methodology that incorperates the notion of Enterprise Ontology. To test the hypothesis, as stated above, the DEMO methodology will be assessed for the purpose of developing QMSs, both theoretically and practically.

3   DEMO

Enterprise Ontology is defined as the implementation independent understanding of the operation of organizations, where organizations are systems in the category of social systems. Being a social system means that the elements are social individuals. One might expect that applying the notion of enterprise ontology contributes to the benefits of applying the institutional paradigm to QMS as explained in Section 2. Since an ontological model abstracts from all implementation and realization issues, the resulting model will not be sensitive to minor process changes (feature 5). As already mentioned, Enterprise Ontology focusses on the interaction between and the responsibilities of social individuals. This facilitates the making unambiguously clear to employees what the authority and responsibility of each of them is. Practice experience has taught that this contributes to cooperation improvement. DEMO [5] incorporates the notion of Enterprise Ontology completely. It claims that the ontological model of an enterprise satisfies the C4E quality criteria [5]. It is:

Coherent: It constitutes a whole.
Consistent: It does not contain any logical contradictions.
Comprehensive: It includes all relevant elements.
Concise: It is kept as brief as possible.
Essential: It is independent of realization and implementation.

At first sight, it seems that the C4E quality requirements are a good match to the desired features 1 to 5. Process description should be kept as concise as possible (feature 1) while it includes all relevant elements (comprehensive)(feature 2). Furthermore the processes should not contain logical contradictions (consistent) and constite a whole (coherent)(feature 3 and 4). Besides these properties, it is very desirable that the document need not be changed for every minor change in the enterprise.Thus, process descriptions should be implementation and realization independent (essential)(feature 5). It can be concluded that Enterprise Ontology is a good fit to the institutional paradigm. As mentioned in Section 2 DEMO is currently the only methodology that incorperates the notion of Enterprise Ontology. Figure 1 shows DEMO's ontological aspect models [5]. The aspect

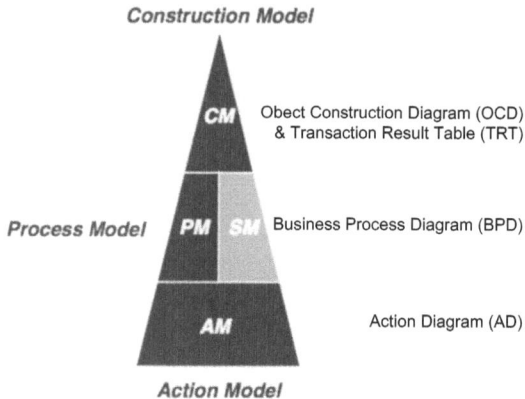

Fig. 1. The ontological DEMO aspect models and the used representations

models can be represented either in diagrams or tables or a combination of the two. Research on representations has turned out that the representations given on the right side of Figure 1 appear practically most appropriate to describe the processes in the QMS handbook [24]. The red parts show the Construction Model (CM), Process Model (PM) and the Action Model (AM). These are the most important models for making QMS handbooks. The pink colored part shows the State Model (SM), which is not used in handbooks. The Organization Construction Diagram (OCD) and Transaction Result Table (TRT) represent the CM. The Business Process Diagram (BPD) represents the PM and the Action Diagram (AD) represents the AM. Each representation will be elaborated by means of a small case in the coming subsections. This case is purely meant to illustrate because the complete model would overshoot the mark of this paper. A fully worked out case can be found in [24].

4   QMS Handbook

4.1   Layout

Determination of the optimal layout is a main topic in the research of Geskus [24]. Figure 2 shows the page layout of the QMS. The first 6 pages contains general information of the regarding enterprise. This is mandatory according the ISO 9001 standard. The last pages in the handbook elaborates the legend of the used symbols besides that a cross reference is included. The focus in this paper is on the process descriptions and therefore are those pages not taken into account in Figure 2. The variable page number 'n' corresponds with the number of processes that are to be described (n starts with 0 and adds up with multiples of 2).

Figure 2 shows that page 7 and 8 are filled with the Actor Description Table (ADT). This table enumerates the identified actor roles with their responsibilities

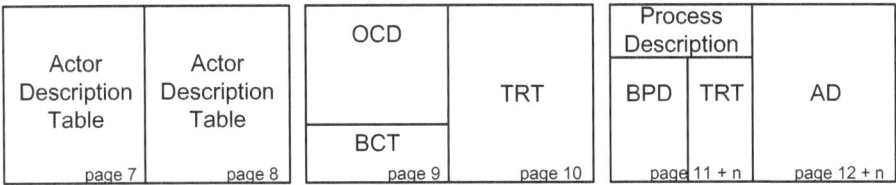

Fig. 2. Layout of the QMS handbook

and authorities. Furthermore, it shows by which function(s) the actor role is fulfilled. The ADT will be elaborated further in subsection 4.2. Page 9 contains the CM represented in the OCD and Bank Contents Table (BCT). The BCT shows information of the external banks. The Transaction Result Table (TRT) is given on page 10, it specifies the results of the identified transactions. The representations of the CM are further elaborated on in subsection 4.3. Each process description takes at least 2 pages where on the left side the textual process description is given. Herein are the aim, required instruments and KPI's of the process defined. Below, the BPD, and the transaction results of the process are given. The elaborated BPD of the case is to be found in 4.4. The AD of the process is given on the right side. The AD is to be found in 4.5. In theory the AD contains all information that is given in BPD so one could say that the BPD is redundant and be omitted. The reason why this is not the case will become clear in subsection 4.4.

4.2 Actor Description Table

In Table 1 the actors of the OCD are listed. Only the internal actors are included. In row BA-01, the order deliverer is marked in the column Managing Partner (MP). This is to be explained as follows: the position MP fulfills the role of order deliverer and is assigned to the responsibilities as listed in the last column. In general a position fulfills at least one actor role and an actor role is fulfilled by at least one position.

Table 1. The Actor Description Table

Nr.	Name	MP	Consultant	Office manager	Responsibilities & Authorities
B-A01	order deliverer	X			• Guarantees that INQA keeps its promises to the customer and guarantees the order quality. • Clear communication between order producer and customer about the order results and the execution of the order. • Is aware of all possible signals indicating a customer's dissatisfaction.
B-A02	order producer	X	X		• Execution of the order. • Share knowledge, expertise and tangible products to reach an optimal result. • Execution of the order according to plan of action.

Practical reflection. Practical experience has revealed that the ADT, particularly on the work floor, brings a significant contribution to the support of the QMS by employees. This first table in the handbook make the employees recognize their operation and the corresponding responsibilities and authorities. This fits directly to the recommendations of Boiral [3]; recognition of the employee in models leads to an increase in support of the QMS among the employees. This support is crucial for a successfulimplementation.

4.3 Representation of the CM

As one can see, the OCD in Figure 3 clearly outlines the scope of the QMS. The Service delivery company has responsibility for all processes within this outline titled: "Service delivery company". These outlines are called boundaries in the DEMO. At the edge of this outline the interactions with the gray colored external actors are shown. E.g: the customer (CA01) requests the delivery of an order from the internal order deliverer B-A01, making it clear that the responsibility of the order delivery lies with the service delivery company.

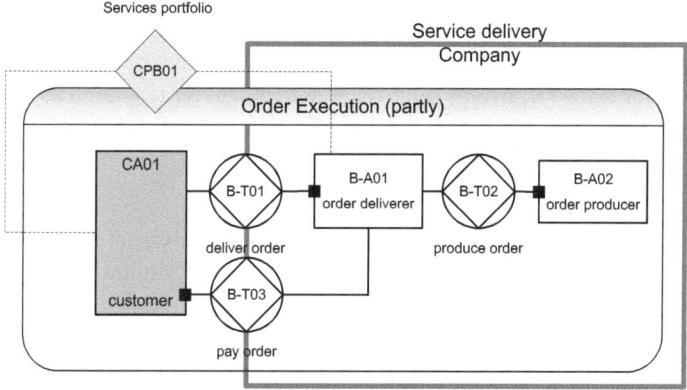

Fig. 3. The Organization Construction Diagram

Transaction Result Table. The transaction results as formulated in [5] are hard to explain to non-DEMO experts when the SM is not given, as is the case in the QMS handbook. Therefore a different notation is chosen to formulate the transaction results. The transaction results are formulated according to the Semantics of Business Vocabulary and Business Rules (SBVR) [25] to distinguish between types and instances. Both notations of the transaction result of B-T01 are given in Table 2. At first sight, one could say that there are not many differences, but it is important to distinguish clearly between type and instance. In DEMO-2 notation it is clear that O is a variable and Order is the type that can be instantiated, e.g.: *Order is delivered* is the type and the instance can be *Order INQ368 is delivered.* In SVBR notation the type always starts with a capital, which does

Table 2. Transaction Results formulated according DEMO-2 and SBVR

Nr.	Transaction name	Nr.	Transaction result(DEMO-2)	Transaction result(SBVR)
B-T01	deliver order	B-R01	Order O is delivered.	Order is delivered.

not look strange to a non-DEMO expert. To avoid unnecessary discussion with non-DEMO experts we have chosen to use the SBVR notation in the handbook. In Table 3 the transaction results of the three transactions are given.

Table 3. Transaction Result Table

Nr.	Transaction name	Nr.	Transaction result
B-T01	deliver order	B-R01	Order is delivered.
B-T02	produce order	B-R02	Order is produced.
B-T03	pay order	B-R03	Order is paid.

Bank Contents Table. Table 4 specifies the contents of the external banks. The dotted lines from an actor role to an external bank are information links. An information link is to be explained as follows: an actor needs external information from the bank in order to do its work. For example, in Figure 3, both the order deliverer and customer are linked to the services portfolio bank: both need to know what kind of services INQA provides.

Table 4. Bank Contents Table

Bank nr.	Bank name	Bank content	Actor nr.	Actor name
B-CPB01	Services portfolio	Services	B-CA02	capacity/knowledge deliverer
		General conditions	B-A01	order deliverer

Practical Reflection. The CM is the most abstract aspect model of DEMO that gives an overview of all identified transactions and actor roles. This is the first model that is shown in the handbook. The CM has proven to be very useful, especially in discussions with managers. This can be explained as follows: the CM is capable, because of its concise and comprehensiveness, to show the essence of an entire enterprise on one page. This powerful instrument function as a strong fundament in discussions with managers about responsibilities and authorities, futhermore it obliges that the discussion is kept on the correct abstraction level. The CM replaces the EFQM Excellence model [26] is a predefined management system that is used to structure an ISO 9001 handbook. It may be clear that the CM is much more useful compared to the EFQM Excellence model because it reflects the true Enterprise.

4.4   Representation of the Process Model

The PM is represented by the BPD. The BPD enables to reveal dependency and causal relations between transactions in a concise way. DEMO experts are able

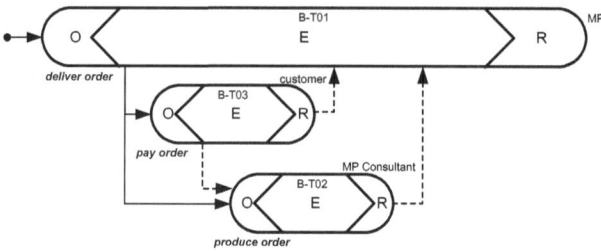

Fig. 4. The Business Process Diagram

to see (Figure 4) in one glance that both B-T02 and B-T03 are component transactions of B-T01. In other words: transactions B-T02 and B-T03 are required to
end-up in the accepted state before B-T01 can be finish the execution step. Besides it shows that the order payment must be promised before the transaction:
produce order can be requested.

Practical Reflection. Although the BPD is a very powerful representation that
enables to show the structure of transactions, practical experience has taught
that the BPD has initially a deterrence effect to non-DEMO experts. The main
cause for this is that the diagram and the used symbols itself are not self-
explanatory, therefore it is chosen to combine the BPD and AD next to each
other as shown in subsection 4.1. Interviews with employees have revealed that
the BPD is skipped at the first moment and the focus is on the AD. After reading the AD by employees, the relation between the AD and BPD becomes clear
and the addition of the BPD becomes evident.

4.5   Representation of the Action Model

The AM is represented by the AD. The AD deviates from the developed pseudo
algorithmic representation of Dietz. The main reason for this is that practical
experience in making DEMO models has taught that the pseudo algorithmic
representation [5] of the AM is hard to interpret by employees without a (computer) science related background. As an alternative the AD has been developed.
Figure 5 shows a single AD element.

C-A0x Fulfiller of role CA0x	**Request** B-T0x **deliver order**	Data carrier:
B-A0x	**Promise** B-T0x **deliver order** **IF <condition>** **ELSE**	
Fulfiller of role B-A0x	Infological Action #1 performed when B-T0x is promised	Data carrier that is needed to perform infological action #1
	Infological Action #2 performed when B-T0x is promised	
	Datalogical Action #1 performed when B-T0x is promised	

Fig. 5. The Action Diagram element

As one can see in Figure 5, the element is build up in 3 columns. The cells in the first column are filled with the actor role. Its fulfiller that correspond with the performed transaction step in the second column. The fulfiller of the actor role is given mainly to lower the understanding threshold that was explained in Section 2. The cells in the second column do always contain a transaction step (e.g. request, promise, etc.) and, if applicable, a condition that needs to be satisfied. Furthermore infological and datalogical actions can be defined. The third column contains the needed data carrier for the condition or to perform the infological or datalogical action. Figure 6 shows the elaborated AD of the service delivery company's order execution process (partly) which is built from AD elements as depicted in Figure 5. As one can see, the actions and conditions inside the AD elements are written in natural language.

Practical Reflection. The AD is the most exhaustive diagram of DEMO. It contains all information that is also contained in the CM and PM; but in a

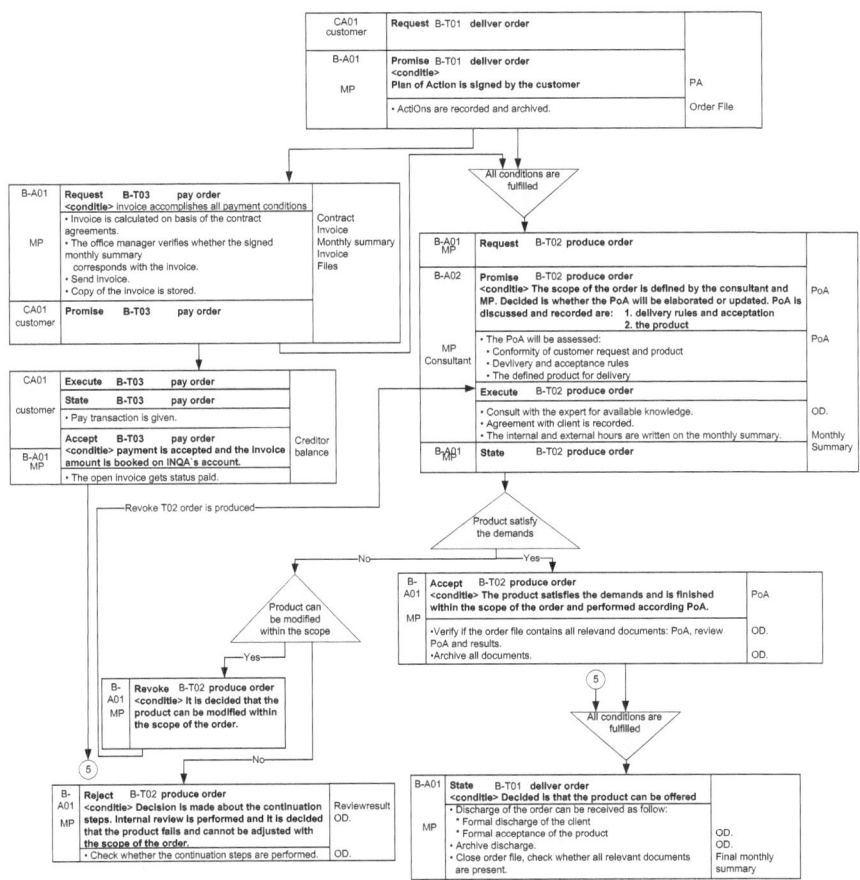

Fig. 6. The Action Diagram

different, and not so easily accessible, way. Practical experience has revealed that the AD is very easy to interpret. This is not very surprising because the AD has a lot of similarities with traditional flowcharts. A disadvantage of the AD might loose its clear overview when the number of relationships between the transaction steps increases. The challenge to optimize the composition in order to realize a clear overview can in result in a tough and time consuming job for the modeler.

5   Evaluation and Conclusions

The new shape of the QMS handbook, as proposed in Section 4, has been tested in two SME companies, namely INQA Quality Consultants and KIPP en Zonen. In both cases the handbook was produced for the purpose of getting the ISO certification. INQA is a service delivery company that conducts ISO certifcations in other companies, Kipp en Zonen produces state-of-the-art high quality electronical devices. In both cases the experiment has been extensively discussed an evaluated. One of the results is that the use of the ADT offers a significant contribution to the support by the employees of the developed QMS. Another one is that the CM has proven to be a powerful instrument in discussions with managers. In particular, the CM has been discussed as a potential substitute for the EFQM Excellence model [26] that was used mostly before. Managers consider the CM to be a substantial improvement compared to the EFQM Excellence model because it represents for them the 'real' enterprise. Lastly, as elaborated in Section 4.4 and 4.5, combining the BPD and the AD leads to an increase in understanding the structure of processes by employees. In addition, the employees get a more clear picture of their responsibilities. We have discussed the results of our approach with the managment of INQA. This has lead to the decision by the management to adopt the new approach in future certification projects. Although the practical evaluation is only based on two cases, the results so far are very interesting and promising. It endorses Boiral's conclusions [3] that the application of a methodology that supports the institutional paradigm results in an increase of commitment by the employees. This has been observed convincingly. Regarding the desirable features of a QMS handbook, as presented in the introduction, the next conclusions can be drawn. First, the produced handbooks are absolutely concise. The reduction in size is estimated at about 80 percent. Next, since this reduction stems mainly from focussing on the ontological model of the enterprise, the other features are achieved as a more or less logical consequence of this. Whether the process descriptions are indeed insensitive to minor changes needs to be experienced yet. However, the evaluations of over 100 other projects in which DEMO has been applied [27] indicate that this will certainly be the case.

References

1. Nederlands Normalisatie-instituut. Nederlandse norm NEN-EN-ISO 9000 (nl). NEN (2000)
2. Aarts, W.M.: Werken met ISO 9001:2000. Samsom (2000)

3. Boiral, O.: Iso 9000: Outside the iron cage. Organization Science 14(6), 720–737 (2003)
4. Dutton, J.E., Dukerich, J.M.: Keeping an eye on the mirror: Image and identity in organizational adaptation. The Academy of Management Journal 34(3), 517–554 (1991)
5. Dietz, J.L.G.: Enterprise Ontology Theory and Methodology. Springer, Heidelberg (2006)
6. Dietz, J.L.G., Hoogervorst, J.A.P.: Enterprise architecture in enterprise engineering. Enterprise Modelling and Information Systems Architecture 3(1c) (2008)
7. Cooperation & Interoperability Architecture & Ontology, http://www.ciaonetwork.org
8. Acharya, U.H., Ray, S.: Iso 9000 certification in indian industries: a survey. Total Quality Management & Business Excellence 11(3), 261–266 (2000)
9. Hughes, T., Williams, T., Ryall, P.: It is not what you achieve it is the way you achieve it. Total Quality Management & Business Excellence 11(3), 329–340 (2000)
10. Dowen, R.J., Docking, D.S.: Market interpretation of ISO 9000 registration. SSRN eLibrary (1999)
11. Carlsson, D., Carlsson, M.: Experiences of implementing ISO 9000 in swedish industry. International Journal of Quality & Reliability Management 13, 36–47 (1996)
12. Scott, R.W.: The adolescence of institutional theory. Administrative Science Quarterly 32(4), 493–511 (1987)
13. Bjorck, F.: Institutional theory: a new perspective for research into is/it security in organisations. In: Proceedings of the 37th Annual Hawaii International Conference on System Sciences, 2004, p. 5 (January 2004)
14. Burrell, G.: Sociological paradigms and organisational analysis: elements of the sociology of corporate life, Arena, Aldershot, Hampshire, pp. 403–426 (1994)
15. Clegg, S.: Power, rule, and domination: A critical and empirical understanding of power in sociological theory and organizational life. International library of sociology. Routledge & Paul (November 1975)
16. Alvesson, M., Willmott, H.: On the idea of emancipation in management and organization studies. The Academy of Management Review 17(3), 432–464 (1992)
17. Parker, M.: Critique in the name of what? postmodernism and critical approaches to organization. Organization Studies 16(4), 553–564 (1995)
18. Weber, M.: The Protestant Ethic and the ”spirit” of Capitalism and Other Writings. Scribners (1958)
19. Cochoy, F., Garel, J.-P., de Terssac, G.: Comment l'ecrit travaille l'organisation: le cas des normes ISO 9000. Revue française de sociologie 39(4), 673–699 (1998)
20. Mispelblom, F.: au-delá de la qualité: Démarches qualité, conditions de travail et politiques du bonheur, Syros, Paris, France (1995)
21. Seddon, J.: The case against ISO 9000. Oak Tree Press, Dublin (2000)
22. Beattie, K.R.: Implementing ISO 9000: A study of its benefits among australian organizations. Total Quality Management & Business Excellence 10(1), 95–106 (1999)
23. Anderson, S.W.: Why firms seek ISO 9000 certification: Regulatory compliance. Production Oper. Management 8(1), 28–43 (1999)
24. Geskus, J.: Demo applied to quality management systems. Master's thesis, Delft University of Technology (2008)

25. OMG. Semantics of business vocabulary and business rules(sbvr), v1.0. In: Semantics of Business Vocabulary and Business Rules(SBVR), v1.0 (2008)
26. Russell, S.: ISO 9000:2000 and the efqm excellence model: competition or cooperation? Total Quality Management 11, 657–665 (2000)
27. Mulder, J.B.F., Dumay, M., Dietz, J.L.G.: Demo or practice: Critical analysis of the language/action perspective. In: Proceedings of the LAP Conference (October 2005)

Analyzing Organizational Structures Using Social Network Analysis

Chuanlei Zhang[1], William B. Hurst[1], Rathinasamy B. Lenin[2], Nurcan Yuruk[1], and Srini Ramaswamy[2]

[1] Department of Applied Science, University of Arkansas at Little Rock,
2801 S. University Avenue, Little Rock, Arkansas 72204, USA
{cxzhang,wbhurst,nxyuruk}@ualr.edu
[2] Department of Computer Science, University of Arkansas at Little Rock,
2801 S. University Avenue, Little Rock, Arkansas 72204, USA
rblenin@ualr.edu, srini@acm.org

Abstract. Technological changes have aided modern companies to gather enormous amounts of data electronically. The availability of electronic data has exploded within the past decade as communication technologies and storage capacities have grown tremendously. The need to analyze this collected data for creating business intelligence and value continues to grow rapidly as more and more apparently unbiased information can be extracted from these data sets. In this paper we focus in particular, on email corpuses, from which a great deal of information can be discerned about organization structure and their unique cultures. We hypothesize that a broad based analysis of information exchanges (ex. emails) among a company's employees could give us deep information about their respective roles within the organization, thereby revealing hidden organizational structures that hold immense intrinsic value. Enron email corpus is used as a case study to predict the unknown status of Enron employees and identify homogeneous groups of employees and hierarchy among them within Enron organization. We achieve this by using classification and cluster techniques. As a part of this work, we have also developed a web-based graphical user interface to work with feature extraction and composition.

Keywords: Business intelligence, organizational hierarchies, classification, clustering, Enron email corpus.

1 Introduction

Technological changes have aided modern companies to gather enormous amounts of data electronically. In this paper we focus in particular, on email corpuses, from which a great deal of information can be discerned about organization structure and their unique cultures. It can also be used as a 'window' by entities such as private equity firms, insurance companies, or banks that aspire to do due diligence in understanding a company's culture, be it in legal, regulatory or other needs. We hypothesize that a broad based analysis of information exchanges (emails in this instance)

A. Albani, J. Barjis, and J.L.G. Dietz (Eds.): CIAO!/EOMAS 2009, LNBIP 34, pp. 143–156, 2009.

among a company's employees could give us deep information about their respective roles within the organization, thereby revealing hidden organizational structures that hold immense intrinsic value. Therefore, these corpuses can be considered as a crucial value-added proposition, which helps one decipher the emerging social structures, as well as potentially identify key personnel (rising stars) within an organization. Such analysis is known as Social Network Analysis (SNA). From a purely business perspective, such type of analysis helps us avoid bias in judging / evaluating the importance of each and every player (employee) of an organization, and providing a detailed view that does not solely depend on key administrative decision makers in modern day hierarchical organizational structures. As a preliminary study, we have earlier used SNA in analyzing software developer roles open-source software (OSS) development to identify key developers and coordinators in making OSS systems more successful [1].

The analysis of social networks has focused on a set of important, but relatively simple measures of network structure [2]; these include issues such as degree distributions, degree correlations, centrality indices, clustering coefficients, subgraph (motif) frequencies, preferential attachment, node duplications, degree distributions and correlations. Recently researchers have begun studying wider community structures in networks and issues such as interconnectedness, empirical relationships, weak community links, collaboration, modularity and community structures [3]. SNA in electronic media essentially involves "Link Mining". Link mining is a set of techniques which is used to model a linked domain using different types of network indicators [4]. A recent survey on link mining can be found in [5]. Its applications include NASDAQ surveillance [6], money laundering [7], crime detection [8], and telephone fraud detection [9]. In [10], the authors showed that customer modeling is a special case of link mining.

The public availability of Enron Corporation's email collection, released during the judicial proceedings against this corporation, provides a real rich dataset for research [11, 12]. In [13, 14], the authors used Natural Language Processing techniques to explore this email data. In [15], the authors used SNA to extract properties of the Enron network and identified the key players during the time of Enron's crisis. In [16], the authors analyzed different hierarchical levels of Enron employees and studied the patterns of communication of the employees among these hierarchies. In [17], the authors used a thread analysis to find out employees' responsiveness. In [18], the authors used an entropy model to identify the most relevant people. In [19], the authors proposed a method for identity resolution in the Enron email dataset. In [20], the authors deployed a cluster ranking algorithm based on the strength of the clusters to this dataset. In [21], the authors provided a novel algorithm for automatically extracting social hierarchy data from electronic communication behavior.

In this paper, we apply SNA to identify different social groups among employees based on the emails they exchanged and attempt to predict organizational structure that emerges from such complex social networks. Such an analysis can be very useful from an economic perspective for reasons such as business strategy, competition, multiplayer perspectives, enabling leadership and innovation support structures, stability and societal benefit. Our work is different from that of most of the earlier works, in that it is significantly more comprehensive. Specifically, it focuses on the following two issues, which are significant value-added propositions for any organization.

1. First, we mine the email corpus to collect the data such as counts of the emails which were exchanged between Enron employees using the To, Cc and Bcc fields, different combinations and ratios of these counts, and the response ratio to emails sent by each employee. The development and use of composite features of these various fields (of email data) in our work is significantly different from all reported earlier work. We use clustering algorithms [22] and classification algorithms [23] to analyze the data in order to identify homogeneous groups of employees and to predict the designation status of employees whose status were undocumented in the Enron database report. For clustering we used matlab and for classification we used weka [24]. Furthermore, we have developed a web-based Graphical User Interface (GUI) that can work with different classifiers to automate feature selection and composition, so that we can interactively carry out the classification analysis. The validity of clusters is demonstrated using different widely used statistical measures [25-28]. The validation of classification-based prediction is done by predicting the designation status of employees, for whom the status is known.

2. Second, we use prediction techniques to identify employees who may be performing roles that are inconsistent with other employees in similar roles within the organization. We hypothesize, that these roles may be associated with either more leadership responsibilities or with more mundane day to day operational responsibilities that keep the organization focused on its core capabilities. Such personnel tend to play either a critically vital role in the organization in helping it accomplish its goals and missions (through leadership, mentoring and training of junior employees) or, are leading indicators of poor performers. This implies that the remaining employees perform within well-defined 'bounded' roles in the organization.

The paper is organized as follows. In section 2, we briefly talk about Enron email corpus and features we extracted from this corpus. In section 3, we discuss about clustering techniques we used and about the GUI we developed to work with weka. In section 4, we discuss the results, and in section 5 we conclude this work and discuss possible future work in this direction.

2   Enron Email Corpus

The Enron email dataset was initially made public by Federal Energy Regulation Commission. The raw dataset can be found at [11]. There are different versions of the datasets processed by different research groups. We collected our dataset from [12], as a MYSQL dump, because the authors cleaned this dataset by removing duplicate emails and processed invalid email address. In this dataset, there are a total of 252759 / 2064442 email messages being sent to / received by a total of 151 Enron employees. Information about each of these employees such as full name, designation, email address, emails sent, emails received, subject and body of emails, and references to these messages is maintained in different tables. For 56 employees, the designation information is unavailable and they are marked as 'N/A' in the dataset. In this paper, we try to establish the designation status of these 56 employees. In Table 1, we tabulate the assigned employee identity numbers, and designation status of 151 employees of Enron [29].

Table 1. Details of Enron employees

ID	Status	ID	Status	ID	Status
1	Director	51	Vice President	101	Director
2	Director	52	Employee	102	N/A
3	Employee	53	CEO	103	Vice President
4	Manager	54	President	104	N/A
5	Employee	55	N/A	105	N/A
6	Employee	56	N/A	106	Director
7	Employee	57	N/A	107	President
8	Employee	58	Trader	108	Manager
9	Employee	59	N/A	109	Manager
10	N/A	60	Trader	110	Managing Director
11	N/A	61	Employee	111	N/A
12	Employee	62	Employee	112	N/A
13	N/A	63	N/A	113	N/A
14	Vice President	64	N/A	114	Employee
15	Employee	65	Managing Director	115	N/A
16	N/A	66	President	116	Employee
17	Manager	67	N/A	117	Manager
18	N/A	68	Vice President	118	Employee
19	Employee	69	Vice President	119	Employee
20	N/A	70	Employee	120	Director
21	Director	71	N/A	121	Director
22	Trader	72	N/A	122	Employee
23	Vice President	73	Employee	123	Manager
24	Manager	74	In House Lawyer	124	Trader
25	Employee	75	N/A	125	Trader
26	Director	76	N/A	126	Vice President
27	Employee	77	Employee	127	CEO
28	N/A	78	Vice President	128	Employee
29	Vice President	79	N/A	129	N/A
30	N/A	80	N/A	130	Director
31	N/A	81	Employee	131	Trader
32	Vice President	82	N/A	132	Trader
33	N/A	83	Vice President	133	Trader
34	N/A	84	N/A	134	Trader
35	Vice President	85	Employee	135	N/A
36	President	86	N/A	136	Employee
37	Vice President	87	Employee	137	Vice President
38	Vice President	88	Trader	138	Trader
39	N/A	89	N/A	139	In House Lawyer
40	Trader	90	Manager	140	Employee
41	N/A	91	N/A	141	N/A
42	Manage	92	Vice President	142	Employee
43	N/A	93	Employee	143	Employee
44	Vice President	94	N/A	144	N/A
45	N/A	95	N/A	145	N/A
46	CEO	96	Vice President	146	N/A
47	Employee	97	N/A	147	In House Lawyer
48	N/A	98	N/A	148	N/A
49	N/A	99	N/A	149	Employee
50	N/A	100	Employee	150	N/A
				151	Employee

The dataset extractions where performed at two different levels of discrimination: 1) Localized Email communications (between Enron employees) and 2) Global Email communications (Email involving Enron employees on a global scale). For the former level, email communications among Enron employees were pulled out of the database. For the latter case, a closer level of discrimination was observed, since an Enron employee could be involved in the Email messages from four different levels of abstraction: To, Cc, Bcc, and From. Scripts (in Perl) were used to connect to the MYSQL database tables and extract the data through table joins to fulfill the data requirements for analysis. Once the data has been extracted out, a secondary set of scripts were used to place the data in formats suitable for the types of analysis planned to be performed. In these final steps the data was placed into tab delimited data files and comma delimited files; with subsequent use either in raw data form, or in matrix type summary formats.

3 Analysis

In this section we discuss briefly the clustering and classification algorithms and features sets we collected from Enron email corpus for these algorithms.

We use classification analysis to predict the designation status of employees whose status are reported as 'N/A' in the email corpus. We use k-means, density based expectation maximization (EM), and tree-random forest techniques for classification analysis [23]. Classification algorithms are based on supervised learning methodology; which assumes the existence of a teacher-fitness function or some other external method of estimating the proposed model. The term "supervised" means "the output values for training samples are known (i.e., provided by a 'teacher')" [30]. In our analysis, we use employee's records for which designation status is known to train the algorithm and use the model obtained from the trained data set to predict the status of employees with 'N/A' values in their designation fields. Validation of prediction using these classification techniques is done using the available designation information of 95 (out of a total of 151) employees, i.e the status of 56 employees was unspecified). We use clustering analysis to identify homogeneous groups, of employees within the Enron organization. To achieve this we use k-means and fuzzy c-means clustering algorithms [22]. Each cluster has a centroid, which is a vector that contains the mean value of each feature for the employees within the cluster. Once clusters are identified, we create a centroid matrix and use hierarchical clustering to identify the similarities and hierarchies among different clusters. It is important to validate cluster results to test whether the right number of clusters is chosen and to test whether cluster shapes correspond to the right groupings in the data.

3.1 Finding the Optimal Number of Clusters

Techniques such as the silhouette measure, partition coefficient, classification entropy, partition index, separation index, and Xie and Beni's index are used to find the optimal number of clusters [25-28]. The silhouette measure is a measure of how close each point in one cluster is to points in the neighboring clusters. The partition coefficient measures the amount of overlap between clusters. Xie and Beni's index quantifies the ratio of the total variation within clusters and separation of clusters. Partition index is the ratio of the sum of compactness and separation of clusters. The classification entropy measures the fuzziness of the cluster partition. Separation index uses a minimum-distance separation for partition validity. The optimal number of clusters is achieved when the first three measures (silhouette measure, partition coefficient, and Xie and Benn index) attain the first local maxima and the later three measures (partition index, classification entropy, and separation index) attain their first local minima.

It is to be noted that not all these techniques are designed to be used with all the chosen clustering techniques. For instance, the silhouette measure is applicable for all clustering techniques whereas partition coefficient and classification entropy are most suitable for fuzzy c-mean algorithm. However, by applying several of these six validation measures together, we can obtain an optimal number of clusters by comparing and contrasting the choices for the number of clusters that concurrently satisfy a majority of these choices.

3.2 Feature Set Identification

For each identifiable employee in the database, we identified nine specific features that could be used for clustering and classification. These features are tabulated in Table 2 and we use these features in our cluster and classification analysis that is reported in section 4. As indicated earlier, the development and use of the ratio features is significantly different from all reported earlier work on the Enron corpus. The major improvement that the use of ratios accomplish is to remove some of the undesirable 'biasing' effects of the raw data collected.

4   Results

In this section, we provide results of our classification and clustering analysis. The consolidated statistics, based on the designation status given in Table 1, is tabulated in Table 3. For our clustering, analysis and prediction we use the groupings as identified in Table 4.

4.1 Classification Analysis Results

For classification analysis, to predict the status of employees with 'N/A' fields, we first tested the dataset using k-means, density based expectation maximization (EM) and tree-random forest techniques. For the analysis we created six different employee groups based on their designation as shown in Table 4 and further assigned numeric ranks to these groups.

We used the features listed in Table 2 for the employees with known designation status as the training data to train the classification algorithms. Steps involved in the process of classification are explained pictorially in Fig. 1. Different techniques (k-means, density based expectation maximization (EM) and tree-random forest) were used to test accuracy. While both k-means and density based EM did not produce an accuracy rate of more than 60%, the tree-random forest technique, produced best

Table 2. Features extracted for Enron email corpus

Feature Id	Feature Description
f_1	Number of emails sent in Cc field
f_2	Number of responses received for emails mentioned for feature f_1
f_3	Cc Response Ratio: f_1 / f_2
f_4	Number of emails sent in To field
f_5	Number of responses received for emails mentioned for feature f_4
f_6	To Response Ratio: f_4 / f_5
f_7	Number of emails received in the To field
f_8	Number of emails received in the Cc field
f_9	Informational Response Ratio: f_7 / f_8

Table 3. Consolidated statistics of Enron employees based on their status

Status	Count
N/A	56
CEO	3
Director	9
Employee	35
In house lawyer	3
Manager	9
Managing Director	2
President	4
Vice President	18
Trader	12

Table 4. Six groups of employees based on their designation

Status	Group Name	Rank
Director, Manager	Middle-Management	3
Vice President, CEO, President	Upper-Management	1
Employee	Employee	4
In House lawyer	In House lawyer	4
Trader	Trader	4
Managing Director	Managing Middle-Management	2

accuracy of 100%. Given the relatively limited size of the data this is expected and hence, the tree-random forest technique is used to predict the 'N/A' status of employees.

Some interesting observations are to be made in the results of predicting employee ranks that were 'N/A'. These are shown in Table 5. In column 2, we predicted the

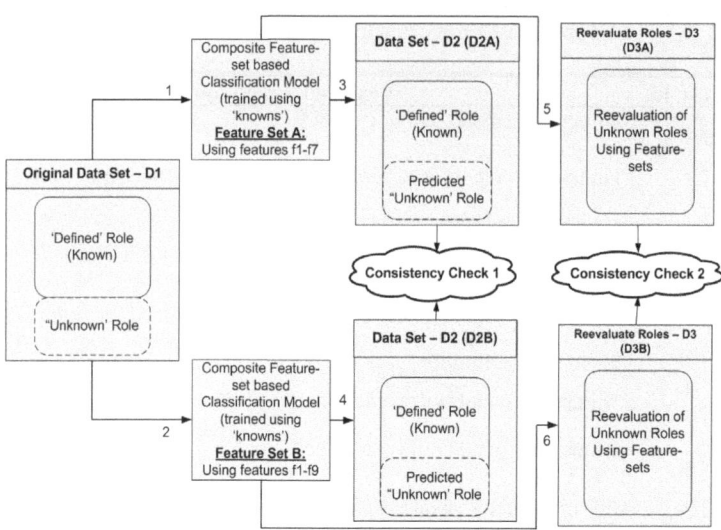

Fig. 1. Steps in classification analysis

rank of 'N/A' employees without using the ratio features f3, f6, and f9 (from table 2). In column 3 of Table 5, the results of predicting the rank of 'N/A' employees without using two key 'differentiation' features – i.e. the feature f8 - where the recipient were sent a 'cc' and the ratio f9, where the ratio of emails sent is computed. In column 4, the prediction of ranks of these 'N/A' employees using all features is presented, while in column 5, we present the results of predicting the roles of all employees – including ranks of those that were previously specified. The objective was to identify and predict some of the subtle differences in the internal roles played within the organizational by these large number of employees that were reported as 'N/A'. From column 5, using the most complete set of features it can be seen that about 41 of these play a upper (23) or middle (18) management ranked employees. Only 15 played the role of an employee (12) or trader (3). This is in stark contrast to not using ratios as features, where 32 of them ranked as either employees (27) or traders (5) in comparison to 23 that ranked as middle (8) or upper management (15) designations.

4.2 Cluster Analysis Results

In this section, we apply cluster analysis for the email dataset to identify homogeneous groups, of employees within the Enron organization. To achieve this we use fuzzy c-means clustering algorithms. We use the silhouette measure, partition coefficient, classification entropy, and Xie and Beni's index, to validate the right number of clusters and fuzzyness of the cluster partition. Once the clusters are identified, we create a centroid matrix and use hierarchical clustering to identify hierarchies among different clusters.

Table 5. Predicting Roles of 'N/A' Designations

Emp ID	Without Ratio	Seven Features	All Features	
			1st round prediction	2nd round prediction
10	Employee	Employee	Upper-Management	Upper-Management
11	Employee	Employee	In-House-Lawyer	Employee
13	Employee	Employee	Upper-Management	Upper-Management
16	Employee	Employee	Upper-Management	Upper-Management
18	Employee	Employee	Upper-Management	Upper-Management
20	Employee	Employee	Employee	Employee
28	Employee	Employee	Middle-Management	Upper-Management
30	Employee	Employee	Upper-Management	Trader
31	Employee	Employee	Upper-Management	Middle-Management
33	Employee	Employee	Employee	Upper-Management
34	Employee	Employee	Managing-Middle-Management	Middle-Management
39	Employee	Employee	Upper-Management	Upper-Management
41	Employee	Employee	Upper-Management	Middle-Management
43	Employee	Employee	Middle-Management	Middle-Management
45	Employee	Employee	Trader	Middle-Management
48	Employee	Employee	Upper-Management	Upper-Management
49	Employee	Employee	In-House-Lawyer	Upper-Management
50	Employee	Employee	Middle-Management	Middle-Management
55	Employee	Employee	Upper-Management	Upper-Management
56	Employee	Employee	Managing-Middle-Management	Middle-Management
57	Employee	Employee	Employee	Employee
59	Employee	Employee	Employee	Employee
63	Employee	In-House-Lawyer	Upper-Management	Upper-Management
64	Employee	Managing-Middle-Management	Trader	Middle-Management
67	Employee	Middle-Management	Upper-Management	Upper-Management
71	Employee	Middle-Management	Middle-Management	Trader
72	Employee	Middle-Management	Employee	Trader
75	In-House-Lawyer	Middle-Management	Upper-Management	Upper-Management
76	Middle-Management	Middle-Management	In-House-Lawyer	Upper-Management
79	Middle-Management	Middle-Management	Employee	Middle-Management
80	Middle-Management	Middle-Management	Upper-Management	Upper-Management
82	Middle-Management	Middle-Management	Managing-Middle-Management	Employee
84	Middle-Management	Trader	Upper-Management	Upper-Management
86	Middle-Management	Trader	Employee	Employee
89	Middle-Management	Trader	Upper-Management	Upper-Management
91	Middle-Management	Upper-Management	Middle-Management	Upper-Management
94	Trader	Upper-Management	Trader	Middle-Management
95	Trader	Upper-Management	Employee	Upper-Management
97	Trader	Upper-Management	Middle-Management	Middle-Management
98	Trader	Upper-Management	Managing-Middle-Management	Upper-Management
99	Trader	Upper-Management	Trader	Employee
102	Upper-Management	Upper-Management	In-House-Lawyer	Middle-Management
104	Upper-Management	Upper-Management	Upper-Management	Middle-Management
105	Upper-Management	Upper-Management	Trader	Employee
111	Upper-Management	Upper-Management	Employee	Upper-Management
112	Upper-Management	Upper-Management	Middle-Management	Middle-Management
113	Upper-Management	Upper-Management	Middle-Management	Upper-Management
115	Upper-Management	Upper-Management	Employee	Employee
129	Upper-Management	Upper-Management	Managing-Middle-Management	Middle-Management
135	Upper-Management	Upper-Management	Upper-Management	Upper-Management
141	Upper-Management	Upper-Management	Upper-Management	Employee
144	Upper-Management	Upper-Management	Employee	Employee
145	Upper-Management	Upper-Management	Middle-Management	Middle-Management
146	Upper-Management	Upper-Management	Middle-Management	Middle-Management
148	Upper-Management	Upper-Management	Middle-Management	Middle-Management
150	Upper-Management	Upper-Management	Employee	Employee

Since the dataset is limited and the expected number of clusters is not too large, we use the fuzzy c-means clustering algorithm to identify the homogeneous group of employees using the features given in Table 2. Based on the validity measures shown in Fig. 2, the optimal cluster size is found to be 6, and these clusters are displayed in Fig. 3. For these 6 clusters, the hierarchy tree is shown in Fig. 4. Clusters 2 and 6 are in the same (bottommost) level, cluster 4 is in the next level, clusters 1 and 5 are in the next level, and cluster 3 is in the top level. The tree structure can be interpreted as follows based on the number of emails exchanged among them: Members of clusters 2 and 6 possess similar characteristics and report to members of cluster 4 who in turn

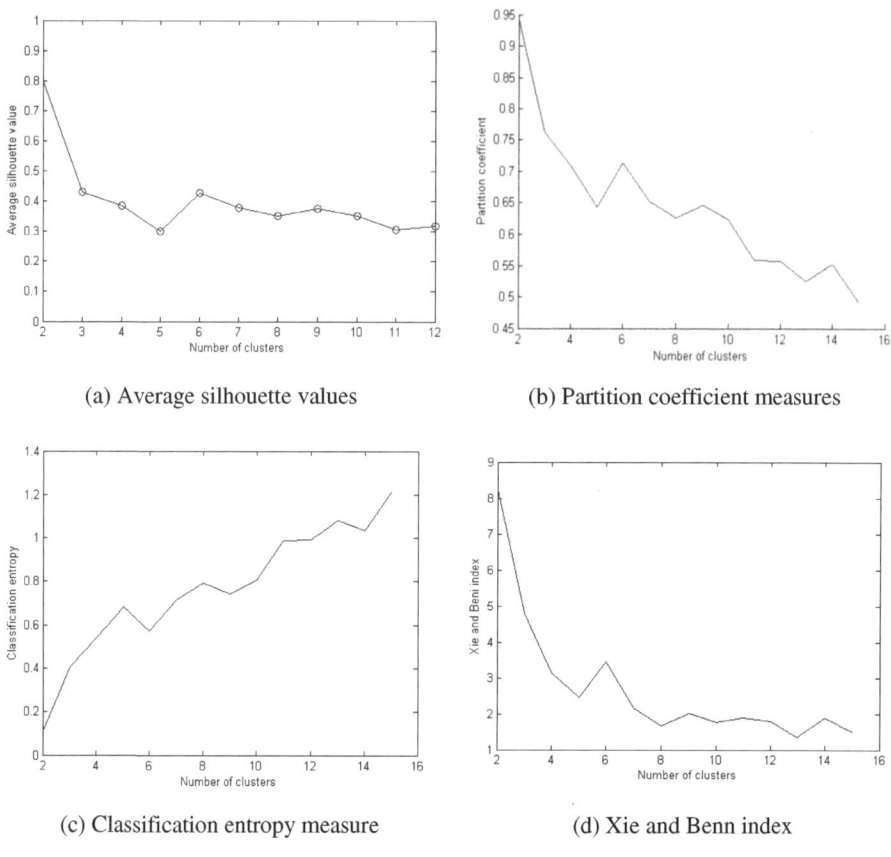

(a) Average silhouette values

(b) Partition coefficient measures

(c) Classification entropy measure

(d) Xie and Benn index

Fig. 2. Validation for optimal number of clusters for fuzzy c-means clustering technique

report to members of cluster 3. Strangely though clusters 1 and 5 possessed similar characteristics, they stay independent of other clusters.

The topmost cluster in the tree is 3 and whose members' ids and status are tabulated in Table 6. For the predicted status, the status group based on Table 4 is provided. Clusters 1 and 5 are isolated from other four clusters. The members of cluster 1 are 75 and 107, and members of cluster 5 are 48, 67, 69, and 73. Except 73 and 75 (employee and middle management), all other members belong to upper management group. We know that member 107 is the President and its interesting that grouped with the president is employee 75 (originally N/A). This person (75) can be considered as a key connection point for other clusters. It is also clear that the management groups are isolated from other groups (clusters 2, 3, 4 and 6) which may be an ideal case for any organization. Similarly, the employee member 73 belongs to cluster 5; which is dominated by upper-management members and function closely with President's group and hence he/she can be considered to be a key connection between cluster 5 and other clusters.

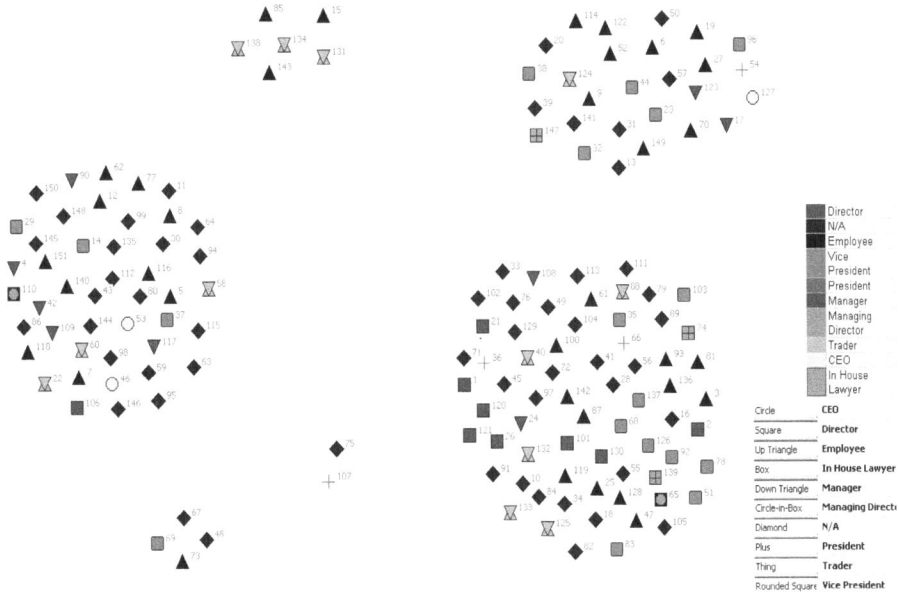

Fig. 3. Identification of employee clusters

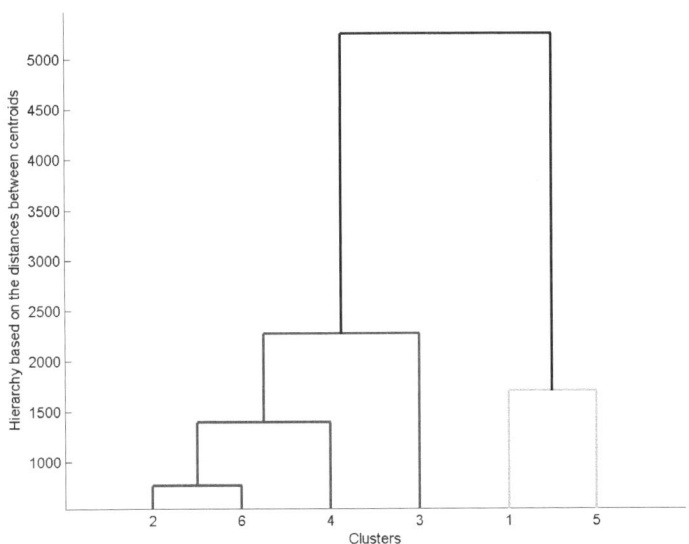

Fig. 4. Hierarchy among clusters (using fuzzy c-means clustering)

The members of cluster 3 (top left in Fig.2) mostly consists of traders and employees. Members in clusters 2, 4 and 6 (top right, bottom right and middle left in Fig. 2) are mixed with members of well-mixed status and hence nothing much can be inferred from this, although members of 2 and 6 work in the same hierarchy level (Fig. 3).

Table 6. Members of top level cluster in the hierarchical tree based on fuzzy c-means clustering technique

ID	Status	Predicted Status (Table 5)
18	N/A	N/A → Upper Management
23	Vice President	*
48	N/A	N/A → Upper Management
67	N/A	N/A → Upper Management
69	Vice President	*
73	Employee	*
75	N/A	N/A → Upper Management
107	President	*
114	Employee	*
137	Vice President	*
147	In House Lawyer	*

Since most of the upper management members are isolated well in clusters 1 and 5 and seem to have acted independent of the other clusters, we can safely conclude that SNA based on the features we extracted from the email corpus verifies our hypothesis: "A broad based analysis of information exchanges (emails in this instance) among a company's employees could give us deep information about their respective roles within the organization, thereby revealing hidden organizational structures that hold immense intrinsic value." In case of Enron, this indicates that the regular employees were probably quite unaware of the emerging problems within the organization. Furthermore, this email corpus also indicates that the presence of management personnel (upper/middle management) personnel who wielded quite some influence, yet performed several 'undefined' roles (N/A's) within the organization.

5   Conclusion

In this paper we carried out a case study of social network analysis on Enron email corpus. Out of 151 employees, there were 56 employees whose status was not reported in the Enron email corpus. As a first step in our analysis, we extracted 9 features from the email corpus. We used these features to predict the unknown status of the employees using the tree random forest classification algorithm from Weka. We further predicted how consistent these 51 employees with respect to their designation status. After predicting the unknown status of employees, we identified homogeneous groups of employees and hierarchy among them using the fuzzy c-mean clustering technique. The results of clustering technique supported our hypothesis that a suitable SNA on electronic data would reveal enough information about strengths and weaknesses of organizations and identify potential employees who played crucial roles in the organization.

In later work, we plan to extend the capability of the web-based GUI that currently leads us perform custom feature composition for analysis, to include clustering / classification analysis using alternate techniques. We plan to identify weighted features based on the response time –ex. for feature such as f2, f4; by modeling response time as a power-law distribution so as to assess relative importance of messages. As appropriate, we plan to use global email communications (emails which have been exchanged between Enron employees and known outsiders – for example lawyers / bankers) to identify the roles played by outsiders. We plan to extend such analysis to study Linux email corpuses to identify key players in the development of Linux, an effort to identify the success behind this open source software system. Such effort is aimed at mimicking identified successes to understand and replicate it for more effective and efficient software development processes.

Acknowledgments

This work is based in part, upon research supported by the National Science Foundation under Grant Nos. CNS-0619069, EPS-0701890 and OISE 0650939 and Acxiom Corporation under contract #281539. Any opinions, findings, and conclusions or recommendations expressed in this material are those of author(s) and do not necessarily reflect the views of the National Science Foundation of Acxiom Corporation.

References

1. Yu, L., Ramaswamy, S., Zhang, C.: Mining email archives and simulating the dynamics of open-source project developer networks. In: Fourth International Workshop on Enterprise and Organizational Modeling and Simulation, Montpellier, France, pp. 17–31 (2008)
2. Wasserman, S., Faust, K.: Social Network Analysis. Cambridge University Press, Cambridge (1994)
3. Wasserman, S., Faust, K.: Social Network Analysis: Methods and Applications. Cambridge University Press, Cambridge (2008)
4. Senator, T.E.: Link mining applications: Progress and challenges. SIGKDD Explorations 7(2), 76–83 (2005)
5. Getoor, L., Diehl, C.P.: Link mining: A survey. SIGKDD Explorations 7(2), 3–12 (2005)
6. Goldberg, H.G., Kirkland, J.D., Lee, D., Shyr, P., Thakkar, D.: The NASD securities observation, news analysis and regulation system (sonar). In: IAAI 2003, pp. 11–18 (2003)
7. Kirkland, J.D., Senator, T.E., Hayden, J.J., Dybala, T., Goldberg, H.G., Shyr, P.: The nasd regulation advanced detection systems (ads). AI Magazine 20(1), 55–67 (1999)
8. Sparrow, M.: The application of network analysis to criminal intelligence: an assessment of the prospects. Social Networks 13, 251–274 (1991)
9. Provost, F., Fawcett, T.: Activity monitoring: noticing interesting changes in behavior. In: Fifth ACM SIGKDD International conference on knowledge discovery and data mining (KDD 1999), pp. 53–62 (1999)
10. Huang, Z., Perlinch, C.: Relational learning for customer relationship management. In: International Workshop on Customer Relationship Management: Data Mining Meets Marketing (2005)
11. Enron, Enron Email Dataset, http://www.cs.cmu.edu/~enron/
12. Adibi, J., Shetty, J.: The Enron email dataset database schema and brief statistical report, Information Sciences Institute (2004)
13. Yang, Y., Klimt, B.: The enron corpus: A new dataset for email classification research. In: European Conference on Machine Learning, Pisa, Italy (2004)

14. McCallum, A., Corrada-Emmanuel, A., Wang, X.: The author-recipient-topic model for topic and role discovery in social networks: Experiments with entron and academic email. In: NIPS 2004 Workshop on Structured Data and Representations in Probabilistic Models for Categorization, Whister, B.C. (2004)
15. Carley, K.M., Diesner, J.: Exploration of communication networks from the enron email corpus. In: Workshop on Link Analysis, Counterterrorism and Security, Newport Beach, CA (2005)
16. Diesner, J., Frantz, T.L., Carley, K.M.: Communication networks from the Enron email corpus. Journal of Computational and Mathematical Organization Theory 11, 201–228 (2005)
17. Varshney, V., Deepak, D.G.: Analysis of Enron email threads and quantification of employee responsiveness. In: Workshop on International Joint Conference on Artificial Intelligence, Hyderabad, India (2007)
18. Adibi, J., Shetty, J.: Discovering important nodes through graph entropy: the case of Enron email database. In: ACM SIGKDD International Conference on Knowledge Discovery and Data Mining, Chicago, Ilinois, U.S.A. (2005)
19. Oard, D.W., Elsayed, T.: Modeling identity in archival collections of email: a preliminary study. In: Third Conference on Email and Anti-spam (CEAS), Mountain View, CA (2006)
20. Bar-Yossef, Z., Guy, I., Lempel, R., Maarek, Y.S., Soroka, V.: Cluster ranking with an application to mining mailbox networks. In: ICDM 2006: Proceedings of the Sixth International Conference on Data Mining, Washington, DC. U.S.A, pp. 63–74 (2006)
21. Rowe, R., Creamer, G., Hershkop, S., Stolfo, S.J.: Automated social hierarchy detection through email network analysis. In: Joint 9th WEBKDD and 1st SNA-KDD Workshop 2007, San Jose, California, USA, pp. 1–9 (2007)
22. Everitt, B.S., Landau, S., Leese, M.: Cluster Analysis, 4th edn. A Hodder Arnold Publication (2001)
23. Izenman, A.J.: Modern Multivariate Statistical Techniques: Regression, Classification, and Manifold Learning, 1st edn. Springer, Berlin (2008)
24. Weka. Weka: Data Mining Software in Java, http://www.cs.waikato.ac.nz/ml/weka/
25. Bensaid, A.M., Hall, L.O., Bezdek, J.C., et al.: Validity-guided (Re)Clustering with applications to image segmentation. IEEE Transactions on Fuzzy Systems 4, 112–123 (1996)
26. Bezdek, J.C.: Pattern Recognition with Fuzzy Objective Function Algorithms. Plenum press (1981)
27. Kaufman, L., Rousseeuw, P.J.: Finding Groups in Data: An Introduction to Cluster Analysis. Wiley, Chichester (1990)
28. Xie, X.L., Beni, G.A.: Validity measure for fuzzy clustering. IEEE Transactions on Pattern Analysis and Machine Intelligence 3(8), 841–846 (1991)
29. Enron Dataset, http://www.isi.edu/~adibi/Enron/Enron.htm
30. Kantardzic, M.: Data Mining: Concepts, Models, Methods, and Algorithms, 1st edn. Wiley/ IEEE (2002)

The Adoption of DEMO: A Research Agenda

Kris Ven and Jan Verelst

Department of Management Information Systems,
University of Antwerp, Antwerp, Belgium
{kris.ven,jan.verelst}@ua.ac.be

Abstract. Organizations are confronted with increasingly complex and dynamic environments. Methodologies in the area of enterprise architecture claim to provide the necessary agility in order to prevail in these environments. The DEMO methodology is one of the few promising methodologies in this area. The need for this type of methodology is now quickly increasing. Research has, however, shown that adoption of methodologies by organizations is often problematic. As a result, notwithstanding their benefits, many methodologies do not diffuse as expected. Moreover, little research has in fact studied the adoption of methodologies. In this paper, we argue that research on the adoption of DEMO is therefore a useful topic that may identify ways to stimulate the acceptance of DEMO. We therefore provide a research agenda on the adoption of DEMO that is firmly grounded in the adoption of innovations literature.

Keywords: enterprise architecture, innovation, adoption, DEMO, enterprise ontology.

1 Introduction

Organizations are confronted with increasingly complex and changing environments. The concept of *Enterprise Architecture* is concerned with providing clear, consistent and coherent design principles for organizational design. Methodologies for enterprise architecture claim to provide the necessary agility in order to prevail in these environments. The DEMO methodology [1, 2] is one of the few promising methodologies in this area. This methodology consolidates over 15 years of theoretical and practical work. As the need for this type of methodology is now quickly increasing, it is interesting to study its adoption by organizations. Therefore, we consider DEMO to be an innovation that is available for organizations to adopt.

Studies have shown that many innovations do not diffuse as expected, irrespective of the benefits they offer [3]. Many factors indeed influence the successful diffusion of innovations across a population of potential adopters. As a result, there is an important gap between the academic world on one hand and practitioners on the other. Studies in the adoption of innovations literature tries to improve upon the evaluation, adoption and implementation of innovations [4]. The importance of adoption research has been acknowledged in several fields.

A. Albani, J. Barjis, and J.L.G. Dietz (Eds.): CIAO!/EOMAS 2009, LNBIP 34, pp. 157–171, 2009.

Although adoption is a key research area within the information systems (IS) literature, it also received attention in fields such as management, marketing, and operations research [3]. Adoption research also has a long history, with the seminal work of Rogers dating back to the early 1960s [5].

Since the DEMO methodology can be used for information systems development (ISD), business process redesign (BPR) and organization engineering (OE), literature on the adoption of IS, BPR and OE methodologies appears relevant. However, given the space available in this paper, we needed to limit our scope. Given the fact that one of the main applications of DEMO appears to be ISD [6], and a strong body of research is available in the IS literature, we will focus our attention on the adoption of IS methodologies. Although many adoption-related studies can be found in IS literature, relatively few of them have focused on the adoption of (IS) methodologies [7, 8]. This is remarkable, given the fact that systems analysis, design and implementation have proven to be problematic in the past 40 years, while the use of methodologies has often been touted as a means to improve upon this situation [9]. The few studies on this topic have shown that the adoption of methodologies is problematic, and that many methodologies do not diffuse widely [10, 9]. It can be expected that DEMO will be subject to the same difficulties as other (IS) methodologies in the past. There are indeed indications that DEMO is only being partially adopted by organizations and that the concept of organization engineering is not fully applied in practice [6]. It is therefore useful to gain more insight into the process by which DEMO is adopted by organizations. To this end, we propose a research agenda on the adoption of DEMO that is firmly grounded in the adoption of innovations literature. It is our expectation that this type of research will assist in the evaluation and adoption of DEMO.

2   Adoption of Innovations

In the literature, many different definitions of an *innovation* are used. Within the context of this paper, we use the most commonly accepted definition, namely to refer to "... *an idea, practice, or object that is perceived as new by an individual or other unit of adoption*" [11, p. 12]. This definition has two important consequences. First, an innovation does not need to be a physical product, and can refer to, for example, a methodology. Second, and more importantly, the term innovation implies that the idea, practice or object must be new to the unit of adoption. This means that it can already be known to other organizations, but that the organization was only exposed to it recently [12]. The term *adoption* is commonly used to refer to the "*decision to make full use of an innovation as the best course of action available*" [11, p. 21]. Hence, adoption refers to the decision of a single actor in a population of interest to use a specific innovation. Literature is also concerned with how the innovation diffuses over this population of interest over time. Diffusion can accordingly be defined as "... *the process by which an innovation is communicated through certain channels over time among the members of a social system*" [11, p. 5].

Significant efforts have been made into studying the adoption behavior of organizations within the IS literature. Most of these studies try to determine the factors that facilitate the adoption and diffusion of IS-related products [3, 13]. The basic research model of IS adoption studies hypothesizes that subjects that have a larger number of favorable characteristics are more likely to exhibit a greater quantity of innovation. The aim of most adoption research is to determine which characteristics influence the adoption of an innovation [3]. It is a firmly established practice in adoption literature to use perceptions of an innovation—so-called secondary attributes [14]—as predictors of usage behavior [11, 15]. This is based on the observation that objective facts about an innovation—so called primary attributes [14]—(e.g., price) can be perceived differently by organizations (e.g., an organization with sufficient resources may consider the price to be acceptable, while an organization with limited resources may consider the price to be too high). This difference in perceptions may lead to different adoption behavior [11].

It is necessary to distinguish between two levels of adoption, namely *individual adoption* and *organizational adoption* [16, 4, 17]. In organizational adoption studies, the unit of analysis is the organization as a whole. The aim of these studies is to determine why decision makers decide to adopt a given technology. In individual adoption studies, the unit of analysis are the employees within the organization. Such studies try to determine why employees decide to use a technology or not, after the organizational adoption decision has been made [16]. Traditionally, the decision to adopt an innovation is first taken by management, after which individuals within the organization will decide on whether to use the innovation. Although adoption of the innovation can be encouraged or even mandated by management, this is not a sufficient condition for successful intraorganizational adoption [7, 9]. Moreover, the managerial decision to adopt an innovation can be based on the suggestion of employees in the organization.

Finally, it must be noted that adoption and non-adoption are two distinct phenomena to study [18, 19]. As noted by Gatignon and Robertson: *"the variables accounting for rejection are somewhat different from those accounting for adoption; rejection is not the mirror image of adoption, but a different form of behavior."* [18, p. 47]. Hence, the absence of any drivers towards the use of an innovation is not the only possible reason for non-adoption. Instead, other factors may limit the tendency of the organization to adopt.

2.1   Organizational Adoption

No universally accepted theory on the organizational adoption of IS currently exists. Already in 1976, Downs and Mohr observed in their review article that much variability in findings in adoption literature exists, which may be caused by the fact that determinants for one innovation do not hold for another class of innovations [14]. Even today, most adoption research focuses on developing theories that are specific to a certain class of technologies and/or contexts [3]. The aim of these studies is to identify a set of factors that influence the organizational adoption decision for a specific technology, or in a specific context.

Nevertheless, there are some frameworks that can be used as a theoretical foundation for studying the organizational adoption decision. A first framework is the *Diffusion of Innovations (DOI)* theory developed by Rogers [11]. According to the DOI theory, the adoption of an innovation is influenced by 5 characteristics of that innovation, namely *compatibility, trialability, observability, relative advantage* and *complexity*. The DOI theory was originally developed to explain the individual adoption of relatively simple innovations. Many authors have pointed out that although DOI has important shortcomings in explaining the organizational adoption decision, it can provide a solid foundation that can be combined with other theories [3].

Another popular framework is the *Technology–Organization–Environment (TOE)* framework. This framework is often used to describe the context in which innovation decision making takes place. According to the framework, there are three elements of an organization's context that influence the decision making process: the *technology context*, the *organizational context* and the *environmental context* [20]. The *technological context* describes the characteristics of existing and new technologies. The *organizational context* refers to a number of descriptive measures of the organization such as size, the degree of formalization and the presence of external linkages (e.g., boundary spanners). The *external context* refers to the environment in which the organization operates, such as the characteristics of the industry and external regulations. TOE does, however, not aim to offer a complete explanation of the organizational adoption decision; it is rather a taxonomy for classifying adoption factors in their respective context. The main contribution of this framework is that it encourages the researcher to take into account the broader context in which innovation takes place, instead of focusing solely on technical characteristics of the innovation.

Results from previous studies in this field have indeed shown that factors related to the innovation itself have a limited impact on the adoption decision. Hence, irrespective of the concrete advantages an innovation may offer, other contextual factors may limit its adoption and diffusion. The importance of taking into account organizational and environmental characteristics in addition to characteristics of the innovation itself, has therefore been supported by various studies [21, 22, 23]. The impact of these organizational and environmental factors can sometimes be quite surprising. For example, since open source software exhibits several unique characteristics (e.g., absence of license costs, the availability of source code, increased trialability), it is commonly expected that these characteristics have an influence on the adoption decision. Nevertheless, a recent study shows that the influence of these technical characteristics have almost no influence on the adoption decision, and that organizational and external factors are much more important [24]. Most adoption studies therefore consider the influence of other contextual factor in the adoption decision.

An important strand in adoption literature conceptualizes the adoption of innovations as the process of *organizational learning* [25, 26, 27]. These scholars argue that in order for organizations to adopt an innovation, they must be able to acquire sufficient information and know-how on that innovation [25, 28]. If organizations have no or insufficient access to information about the innovation, this may become an important barrier for its adoption. The term *absorptive capacity* is used to refer to the degree to which organizations can acquire and process new information about an innovation [26]. Hence, the prior and current knowledge of the organization have an important impact on the innovativeness of the organization [26]. If the required knowledge is not available internally, the organization may rely on external knowledge in the form of consultants and service providers. Another important way to overcome knowledge barriers are *boundary spanners* in the organization. Boundary spanners are individuals in the organization who connect their organization to external information and can bring the organization in contact with new innovations [20, 29]. Boundary spanners are especially important when external knowledge is difficult to obtain by the internal staff [26]. One example in which the role of boundary spanners was very clear, was in the adoption of open source software. It was frequently observed that employees in the organization became familiar with open source software outside their work environment. When a suitable project presented itself within their organization, they would suggest the use of open source software [24]. In this respect, the adoption of open source software is frequently a bottom-up initiative. This suggests that the traditional top-down perspective (organizational adoption precedes individual adoption) may not always hold in practice. In such cases, someone from management with decision authority may in a second phase take on the role of product champion to further promote the use of the innovation [30].

Although many early studies considered adoption to be a binary event, several empirical studies have shown that this conceptualization is a too raw measure. These studies have highlighted the existence of so-called *assimilation gaps* [31]. These assimilation gaps are caused by a difference in time between which the organization decides to adopt (acquire) an innovation and the widespread deployment of that innovation in the organization. Therefore, it has been argued that it is important to obtain more in-depth insight into which *assimilation stage* an organization has progressed [32, 27]. Assimilation can be seen as a process that is set in motion when the organization first becomes aware of an innovation, makes the decision to acquire that innovation and then fully deploys that innovation [32].

2.2 Individual Adoption

In contrast to organizational adoption, a number of generally accepted models exist in literature to explain the individual adoption decision. Although several models exist, the most important ones are the *Theory of Reasoned Action (TRA)* [33], the *Theory of Planned Behavior (TPB)* [34], and the *Technology Acceptance Model (TAM)* [35] (for a complete list, we refer to [17]). Of these models, TAM has gained a dominant position in individual adoption research. It assumes that the decision of an individual on whether to adopt or reject a new technology is directly influenced by the *intention to use*. The intention to use is influenced by the *perceived usefulness* and *perceived ease of use* of the technology. The latter two factors are believed to be two relevant factors shaping the intention of end users to use the technology. The more positive the outcome of interacting with the system is, the higher the tendency to use the system will be. Similarly, if a system is easy to use, the user will believe in his or her ability to use the system. Furthermore, it is hypothesized that improvements in the usability of the software may lead to increased performance. Hence, there is a positive relationship between perceived ease of use and perceived usefulness [36]. Several authors have proposed modifications and extensions to TAM in order to realize a better statistical model fit and increase the variance explained by the model. The two most important extensions are TAM2 [37] and UTAUT [17]. Overall, literature has provided strong empirical support for the validity of TAM.

Despite its broad application in adoption literature, prominent scholars currently question the simple replication of TAM (including some minor extensions to it). Although TAM provides a parsimonious model, it offers limited practically useful information for managers who wish to stimulate the use of an innovation within the organization. In a recent special issue in the *Journal of the Association for Information Systems*, several opportunities are described to heighten the understanding of individual adoption decisions [38]. Another promising direction is the development of TAM3 in which a number of antecedents of perceived usefulness and perceived ease of use are described [39]. Results from applying this model provides managers with more insight into which interventions could lead to a greater acceptance and effective utilization of innovations.

3 Adoption of IS Methodologies

In this section, we focus our attention on the adoption of IS methodologies. Several studies have investigated the adoption of tools (e.g., CASE) or techniques (e.g., object-oriented or component-based development). However, these results are not directly applicable to the adoption of methodologies, since this adoption is much more invasive for an organization. Studies have shown that other determinants are (more) important in predicting the acceptance of a methodology than those influencing the acceptance of tools or techniques [7,9]. Based on the large number of methodologies available—often referred to as the *methodology jungle* [40]—it could be expected that their adoption has formed the topic of intense research. However, relatively few studies have investigated the adoption of methodologies, and the determinants of methodology acceptance in particular [7,8]. An overview of the most relevant studies in the context of this paper is shown in Table 1.

3.1 Organizational Adoption

As can be seen in Table 1, some research has been performed on the organizational adoption of methodologies. These studies were primarily limited to measuring the use of methodologies and their degree of customization [41,10,42], or to investigate the rejection of a methodology in a single organization [43]. Results

Table 1. Exemplary Works on the Adoption of Methodologies

Level	Studies
Organizational adoption	Formalized Systems Development Methodologies [10]
	Systems development methodologies [41]
	Structured systems development methods [42]
	Structured Systems Analysis and Design Method (SSADM) [43]
Individual adoption	Structured Systems Analysis (SSA) [44]
	Custom structured development methodology [7]
	Systems development methodologies [45]
	Custom structured development methodology [9]
	Agile methodologies [46]
	Extreme Programming [47]
	Multiview methodology [48]

show that only 40–44% of organizations are using a formal systems development method [10, 42]. It also appears that developers only use the techniques prescribed by the methodology, and do not absorb its underlying philosophy [10]. Moreover, most organizations (85–94%) tend to customize the methodology to their specific needs and environment [10, 42]. Such use of a methodology may limit the value of the methodology. It has been noted that little is known about the actual success of customizing a methodology [42]. Nevertheless, it appears that deviations from the methodology are in most cases deliberate and the consequences of this action are considered [41]. Organizations also primarily focus on the high-level guidelines provided by the methodology, since the detailed guidelines are not considered useful or applicable [41]. Overall, it appears that methodology usage is more likely in large organizations, large IS departments, organizations with much in-house development and in long projects [10].

Although these studies provide useful insight into the use of methodologies, no prior study—based on the adoption of innovations literature—has attempted to determine the reasons that influence the decision to use a methodology at the organizational level.

3.2 Individual Adoption

At the individual level, several studies have been concerned with identifying the acceptance of methodologies by developers. A summary of the factors that have empirically been shown to be related to methodology acceptance by individual analysts and developers is shown in Table 2. These results show that not only technical factors have an impact on individual acceptance, but also organizational, individual and behavioral factors [43, 44]. Some of these results require some elaboration.

Results show mixed findings for the impact of developer experience on methodology acceptance. Some studies have found a positive effect, arguing that experienced developers are better able to recognize the merits of a methodology [44]. Other studies suggest a negative effect, suggesting that experienced

Table 2. Factors Influencing the Individual Adoption of Methodologies

Level	Studies
Technical	– *Relative advantage/usefulness:* Methodologies that are useful for developers are more easily adopted [44, 7, 9]
Organizational	– *Reward structure:* Using the methodology should be rewarded by the organization (also related to usefulness) [44, 7, 9] – *Time pressure:* Following the methodology could delay a project [44]. – *Budget constraints:* Lower(ing) budgets has a negative impact on use [44]. – *Training and education:* Training and education helps increase use, but should immediately preceed use and is not sufficient [44, 45, 48] – *Supervisor role:* Supervisors can be gatekeepers, stimulating or discouraging use [44, 7, 45, 47] – *Organizational mandate:* Use is stimulated by an organizational mandate (in contrast to voluntary use) [7, 9, 47] – *Client perceptions:* Customer may support or ask the use of a methodology [44] – *Compatibility:* The methodology should be consistent with current work practices [7, 9, 46, 47]
Personal	– *Developer experience:* Conflicting evidence: experience can be positively [44] or negatively [41] related to use. – *Subjective norm:* Advice and opinions from colleagues is very important [44, 7, 9]

developers may feel constrained by the guidelines of a methodology [10]. Based on an in-depth qualitative study, Fitzgerald suggests a U-curve relationship between both variables [41]. According to this curve, methodology usage declines with increasing experience of developers, but starts to rise again when the methodology is heavily customized, thereby meeting the needs and requirements of developers and the organization.

The compatibility of the methodology with current practices was also found to be a very important factor in the acceptance [7, 9, 49, 46]. If the methodology is not compatible with current practices, its use will be very disruptive for developers. As a result, it has been recommended to gradually introduce a methodology in the organization, possibly customizing parts that deviate from current practices [7,9]. This may be one of the reasons why methodologies are frequently customized. There are indications that even agile methodologies—that are known to be more flexible and developer-friendly—tend to be frequently customized [46].

The opinion of colleagues also has an important impact on the use of methodologies [44, 7, 9]. This factor is usually referred to as *subjective norm*, reflecting the idea that developers feel that their colleagues—whose opinion is valued— think that they should use the methodology [37]. In addition, direct supervisors may also have an influence by deciding to stimulate methodology usage or not. Supervisors can therefore become gatekeepers, allowing developers and analysts access to the methodology or not [44, 7, 45, 47].

Other social factors can also be important in the individual acceptance of a methodology. For example, one study found that the use of a methodology can be stimulated by customers who are convinced of its benefits (e.g., higher quality) [44]. It appears, however, that many customers were convinced by developers who were advocating the use of the methodology [44]. It remains to be seen whether similar behavior can be found in other organizations, since previous research has also shown that managers are generally more positive towards the benefits of methodologies than developers [45]. Therefore, several authors have argued that organizational benefits realized by methodology usage should be translated into personal benefits for developers to increase acceptance [44, 7, 9].

3.3 Adoption of DEMO

To our knowledge, only one prior study has investigated the use of DEMO. Based on a survey among 50 subjects, and a panel consisting of 19 experts, the study offered the following useful insights [6]:

- DEMO was used in 43% of the cases for business process redesign (BPR), and in 37% of the cases for information systems development (ISD). DEMO did not appear to be widely used for organization engineering.
- DEMO was most often used in larger groups of developers and analysts.
- DEMO was primarily used by medium-sized and large organizations.
- DEMO was primarily used in short projects (4–6 months).
- Most organizations only used parts of DEMO (e.g., only the process model or information model), supplemented with traditional techniques (e.g., UML or Petri Nets). Which parts of DEMO were used, seemed related to the domain in which DEMO was applied (i.e., BPR or ISD).

Overall, these findings are largely consistent with studies on the adoption of other methodologies. It appears that organizations select those elements from DEMO they deem useful, and further customize the methodology by including other well-known techniques. It is, however, remarkable that DEMO appears to be used primarily for short projects, while its benefits could be expected to be higher in large-scale projects.

4   Proposed Research Agenda

Based on the literature described in the previous two sections, we now present a research agenda on the adoption of DEMO that is grounded in the adoption of innovations literature. Similar to previous studies, we distinguish between research topics at the organizational and individual level.

4.1   Organizational Adoption

Notwithstanding the fact that most studies on the adoption of methodologies have considered the individual adoption decision, we believe that most relevant and interesting topics on the adoption of DEMO could be studied at the organizational level.

First of all, studies could be conducted to identify the reasons that influence the decision of an organization to adopt DEMO. Although this topic has received less attention in the adoption of methodologies literature, it is frequently studied in the adoption of innovations literature (e.g., [24,27,23,22]). Frameworks such as DOI and TOE could act as a theoretical framework for developing a conceptual model describing the factors that influence the organizational adoption decision. The reasons for adoption would go further than just determining the benefits of using DEMO. In addition, similar to individual acceptance [43,44], non-technical factors are frequently found to have an impact on the organizational adoption decision [24, 20]. As previously noted, non-adoption is considered a different research topic. Hence, a similar study could be conducted on determining the barriers to the adoption of DEMO. Both lines of research would provide more insight into the organizational and environmental factors that have an impact on the adoption decision. This would also allow to devise appropriate measures to stimulate the adoption of DEMO, or to identify potential enhancements to DEMO.

Second, it would be interesting to study the adoption of DEMO from an organizational learning perspective. Previous research has, for example, shown that organizations may be faced with considerable *knowledge barriers* when adopting a new innovation [25,26,24,27]. Hence, the adoption behavior of organizations is frequently influenced by the knowledge possessed by the organization, and how easily the organization can acquire new knowledge [25,26]. It would therefore be very interesting to consider how knowledge about DEMO is acquired, managed and disseminated within organizations. For example, the adoption of DEMO could be a bottom-up initiative in which developers or analysts learn about DEMO through external training sessions or seminars, become interested in applying the methodology and try to promote its use within their organization. However, this would require that other members of the organization can be effectively educated about DEMO. The results of this research could provide more insight into how the process of organizational learning could be most effectively managed by organizations. In addition, results may indicate how DEMO is actually introduced in the organization, and by which members of the organization.

Third, research shows that only some parts methodologies are actually adopted by organizations, and others are (heavily) customized. The study of Dietz et al. [6] also showed that organizations did not adopt the whole DEMO methodology. It would therefore be interesting to study which parts of DEMO are actually adopted by organizations (cfr., *way of thinking, way of working, way of managing, way of modeling*). Given the results of previous research [10], it is possible that organizations only adopt some modeling elements of DEMO, but not the *way of thinking* that is fundamental to DEMO. Which elements of DEMO are used, may be related to the area in which DEMO is applied (e.g., information systems development, business process redesign, or organization engineering). This may provide more insight into which elements of DEMO are valued by organizations, and which elements remain unused. The evaluation framework developed by Vavpotic and Bajec [50] appears to be very useful in this regard, as it evaluates methodologies on both *technical suitability* (i.e., efficiency) and *social suitability* (i.e., adoption). Such evaluation may also indicate potential mismatches between the theoretical merits of certain elements of the methodology and how they are perceived by practitioners. This may give impetus to improvement scenarios to stimulate full use of the methodology.

Fourth, research could be conducted on the consequences of using a derived version of DEMO. As previously noted, little is known about whether in-house customizations to methodologies are successful [42]. Research on this topic could provide more insight into the benefits and risks involved in using a customized version of DEMO.

A final research topic concerns the *assimilation gap* theory of Fichman [31]. Research could be conducted to investigate how organizations progress through the various assimilation stages and at what rate. The results would provide more insight into how intensively DEMO is used, and whether it is fully deployed and institutionalized within organizations. The existence of assimilation gaps (i.e., a long period of time between first adoption and full institutionalization) could indicate that organizations are faced with considerable knowledge barriers. Results from this study could provide more insight into which factors can help bridging both the assimilation gap, allowing organizations to fully deploy DEMO.

4.2 Individual Adoption

A first and obvious research topic at the individual adoption level, consists of replicating and adapting previous studies on the adoption of methodologies within the context of the adoption of DEMO. To study individual acceptance, research models specifically developed for studying the adoption of methodologies could be used (e.g., [9]). Alternatively, generally accepted models such as TAM could be used as well. One interesting line of research would be to apply TAM3 [39] to the adoption of DEMO, as this would provide more insight into the antecedents of perceived usefulness and perceived easy of use. This, in turn, would allow for making interventions that could increase the individual acceptance of DEMO.

Second, research could be conducted on the role of compatibility in the adoption of DEMO. Previous research has shown that compatibility of a methodology with current practices is a very strong determinant of acceptance [7,9,46,47]. It is possible that DEMO is rather incompatible with the knowledge and practices of most developers. This could be a considerable challenge for a successful adoption. Therefore, research could be specifically conducted on how organizations can best overcome this issue. Evidently, training developers in DEMO is one important way, but is likely to be insufficient by itself [44,45,48]. Other authors have suggested to follow an incremental approach during which parts of the methodology are gradually introduced in the organization, possibly customizing some elements [44]. Evidently, this research topic may be related to studying the organizational learning process at the organizational adoption level. Hence, this research could result in practical guidelines on how DEMO can be most efficiently adopted by organizations.

Finally, although we referred to users of a methodology in this paper as "developers" or "analysts", another research topic could consist of gaining a deeper insight into who the users of DEMO are within organizations. This would indicate if DEMO is primarily used by developers or analysts, or rather by managers or employees with a more business-oriented profile. It could also investigate if a relationship exists between the profile of DEMO users and the domain in which DEMO is used (e.g., ISD, BPR or OE).

4.3   Prioritization and Concretizing of Research Efforts

Based on our research agenda, we identified three topics that should receive priority. These topics are, in order of importance: *determining the enablers and inhibitors for adoption* (Sec. 4.1, topic 1); *determining which parts of DEMO are being adopted* (Sec. 4.1, topic 2); and *determining the users of DEMO within organizations* (Sec. 4.2, topic 3). Studying these topics will provide more insight into how the adoption of DEMO can be increased, and which potential enhancements to DEMO can be made.

Given the rather exploratory nature of this research topic, one of the main aims will be to formulate and test concrete hypotheses with respect to the adoption of DEMO. Although some of these research topics may seem trivial at first sight, experience shows that providing a well-founded answer is far less evident and rich insights may emerge. The adoption of innovations literature provides a solid and comprehensive framework to support this analysis. Furthermore, it can be combined with other perspectives (e.g., knowledge exchange and change management) to enrich the analysis. Although this paper only discussed the adoption of IS methodologies, it also seems fruitful to consider studies on the adoption of BPR and OE methodologies (e.g., Viable Systems Model and Modern Sociotechnique).

5   Conclusion

The adoption of innovations literature shows that many innovations do not diffuse as expected. The adoption of methodologies in particular appears to be problematic. Surprisingly, relatively few studies have been conducted on this topic. It is likely that DEMO will experience similar difficulties in finding acceptance in organizations to other methodologies. In addition, there are indications that organizations only adopt part of the DEMO methodology. Hence, we strongly believe that the research agenda described in this paper has the potential to provide significant insight into the adoption of DEMO by organizations. This research will also result in concrete interventions that can be undertaken to stimulate the adoption of DEMO. These interventions will be much more fundamental than the *"tips and tricks"* frequently found in practitioner literature (e.g., [49]). It can be expected that many of these interventions will be a reaction to factors in the organizational, environmental or (inter)personal context in which the adoption takes place.

The main contribution of this paper is the derivation of a research agenda on the adoption of DEMO that is strongly grounded in the adoption of innovations literature. The results from the proposed research has the potential to improve upon the evaluation, adoption and implementation of DEMO by organizations.

References

1. Dietz, J.L.: The Deep Structure of Business Processes. Communications of the ACM 49(5), 58–64 (2006)
2. Dietz, J.L.: Enterprise Ontology: Theory and Methodology. Springer, Berlin (2006)
3. Fichman, R.G.: The Diffusion and Assimilation of Information Technology Innovations. In: Zmud, R. (ed.) Framing the Domains of IT Management: Projecting the Future Through the Past, pp. 105–128. Pinnaflex Educational Resources, Cincinnati (2000)
4. Fichman, R.G.: Information Technology Diffusion: A Review of Empirical Research. In: DeGross, J.I., Becker, J.D., Elam, J.J. (eds.) Proceedings of the 13th International Conference on Information Systems (ICIS 1992), Dallas, TX, December 13–16, 1992, pp. 195–206. Association for Information Systems, Atlanta (1992)
5. Rogers, E.M.: Diffusion of Innovations, 1st edn. The Free Press, New York (1962)
6. Dietz, J., Dumay, M., Mulder, H.: Demo or Practice: Critical Analysis of the Language/Action Perspective (2005)
7. Riemenschneider, C.K., Hardgrave, B.C., Davis, F.D.: Explaining Software Developer Acceptance of Methodologies: A Comparison of Five Theoretical Models. IEEE Transactions on Software Engineering 28(12), 1135–1145 (2002)
8. Wynekoop, J.L., Russo, N.L.: Studying System Development Methodologies: an Examination of Research Methods. Information Systems Journal 7(1), 47–65 (2003)
9. Hardgrave, B.C., Davis, F.D., Riemenschneider, C.K.: Investigating Determinants of Software Developers' Intentions to Follow Methodologies. Journal of Management Information Systems 20(1), 123–151 (2003)
10. Fitzgerald, B.: An Empirical Investigation into the Adoption of Systems Development Methodologies. Information & Management 34(6), 317–328 (1998)
11. Rogers, E.M.: Diffusion of Innovations, 5th edn. The Free Press, New York (2003)
12. Daft, R.L.: A Dual-Core Model of Organizational Innovation. Academy of Management Journal 21(2), 193–210 (1978)
13. Swanson, B.E.: Information Systems Innovation among Organizations. Management Science 40(9), 1069–1092 (1994)
14. Downs Jr., G.W., Mohr, L.B.: Conceptual Issues in the Study of Innovation. Administrative Science Quarterly 21(4), 700–714 (1976)
15. Zaltman, G., Duncan, R., Holbek, J.: Innovations and Organizations. John Wiley & Sons, New York (1973)
16. Frambach, R.T., Schillewaert, N.: Organizational Innovation Adoption: A Multilevel Framework of Determinants and Opportunities for Future Research. Journal of Business Research 55(2), 163–176 (2002)
17. Venkatesh, V., Morris, M.G., Davis, G.B., Davis, F.D.: User Acceptance of Information Technology: Toward a Unified View. MIS Quarterly 27(3), 425–478 (2003)

18. Gatignon, H., Robertson, T.: Technology Diffusion: An Empirical Test of Competitive Effects. Journal of Marketing 53(1), 35–49 (1989)
19. Nabih, M.I., Bloem, S.G., Poiesz, T.B.: Conceptual Issues in the Study of Innovation Adoption Behavior. Advances in Consumer Research 24(1), 190–196 (1997)
20. Depietro, R., Wiarda, E., Fleischer, M.: The Context for Change: Organization, Technology and Environment. In: Tornatzky, L.G., Fleischer, M. (eds.) The Processes of Technological Innovation, pp. 151–175. Lexington Books, Lexington (1990)
21. Damanpour, F.: Organizational Innovations: A Meta-analysis Of Effects Of Determinants And Moderators. Academy of Management Journal 34(3), 555–590 (1991)
22. Grover, V., Goslar, M.D.: The Initiation, Adoption, and Implementation of Telecommunications Technologies in U.S. Organizations. Journal of Management Information Systems 10(1), 141–163 (1993)
23. Thong, J.Y.: An Integrated Model of Information Systems Adoption in Small Businesses. Journal of Management Information Systems 15(4), 187–214 (1999)
24. Ven, K., Verelst, J.: The Organizational Adoption of Open Source Server Software: A Quantitative Study. In: Golden, W., Acton, T., Conboy, K., van der Heijden, H., Tuunainen, V. (eds.) Proceedings of the 16th European Conference on Information Systems (ECIS 2008), Galway, Ireland, June 9–11, 2008, pp. 1430–1441 (2008)
25. Attewell, P.: Technology Diffusion and Organizational Learning: The Case of Business Computing. Organization Science 3(1), 1–19 (1992)
26. Cohen, W.M., Levinthal, D.A.: Absorptive Capacity: A New Perspective on Learning and Innovation. Administrative Science Quarterly 35(1), 128–152 (1990)
27. Fichman, R.G., Kemerer, C.F.: The Assimilation of Software Process Innovations: An Organizational Learning Perspective. Management Science 43(10), 1345–1363 (1997)
28. Swanson, E.B., Ramiller, N.C.: Innovating Mindfully with Information Technology. MIS Quarterly 28(4), 553–583 (2004)
29. Tushman, M.L., Scanlan, T.J.: Characteristics and External Orientations of Boundary Spanning Individuals. Academy of Management Journal 24(1), 83–98 (1981)
30. Beath, C.M.: Supporting the Information Technology Champion. MIS Quarterly 15(3), 355–372 (1991)
31. Fichman, R.G., Kemerer, C.F.: The Illusory Diffusion of Innovation: An Examination of Assimilation Gaps. Information Systems Research 10(3), 255–275 (1999)
32. Fichman, R.G.: Going Beyond the Dominant Paradigm for IT Innovation Research: Emerging Concepts and Methods. Journal of the Association for Information Systems 5(8), 314–355 (2004)
33. Fishbein, M., Ajzen, I.: Belief, Attitude, Intention and Behavior: An Introduction to Theory and Research. Addison-Wesley, Reading (1975)
34. Ajzen, I.: The Theory of Planned Behavior. Organizational Behavior and Human Decision Processes 50(2), 179–211 (1991)
35. Davis, F.D.: Perceived Usefulness, Perceived Ease of Use, and User Acceptance of Information Technology. MIS Quarterly 13(3), 319–340 (1989)
36. Davis, F.D., Bagozzi, R.P., Warshaw, P.R.: User Acceptance of Computer Technology: A Comparison of Two Theoretical Models. Management Science 35(8), 982–1003 (1989)
37. Venkatesh, V., Davis, F.D.: A Theoretical Extension of the Technology Acceptance Model: Four Longitudinal Field Studies. Management Science 46(2), 186–204 (2000)

38. Hirschheim, R.: Introduction to the Special Issue on Quo Vadis TAM—Issues and Reflections on Technology Acceptance Research. Journal of the Association for Information Systems 8(4) (2007)

39. Venkatesh, V., Bala, H.: Technology Acceptance Model 3 and a Research Agenda on Interventions. Decision Sciences 39(2), 273–315 (2008)

40. Avison, D.E., Fitzgerald, G.: Information Systems Development: Methodologies, Techniques and Tools, 2nd edn. McGraw-Hill, London (1995)

41. Fitzgerald, B.: The Use of Systems Development Methodologies in Practice: a Field Study. Information Systems Journal 7(3), 201–212 (1997)

42. Hardy, C.J., Thompson, J.B., Edwards, H.M.: The Use, Limitations and Customization of Structured Systems Development Methods in the United Kingdom. Information and Software Technology 37(9), 467–477 (1995)

43. Sauer, C., Lau, C.: Trying to Adopt Systems Development Methodologies—A Case-based Exploration of Business Users' Interests. Information Systems Journal 7(4), 255–275 (1997)

44. Leonard-Barton, D.: Implementing Structured Software Methodologies: A Case of Innovation in Process Technology. Interfaces 17(3), 6–17 (1987)

45. Huisman, M., Iivari, J.: Deployment of Systems Development Methodologies: Perceptual Congruence between IS Managers and Systems Developers. Information & Management 43(1), 29–49 (2006)

46. Mahanti, A.: Challenges in Enterprise Adoption of Agile Methods — A Survey. Journal of Computing and Information Technology 14(3), 197–206 (2006)

47. Toleman, M., Ally, M., Darroch, F.: Aligning Adoption Theory with Agile System Development Methodologies. In: Proceedings of the 8th Pacific-Asia Conference on Information Systems (PACIS 2004), Shanghai, China, July 8–11 (2004)

48. Kautz, K., Pries-Heje, J.: Systems Development Education and Methodology Adoption. SIGCPR Computer Personnel 20(3), 6–26 (1999)

49. Kroll, P., Kruchten, P.: The Rational Unified Process Made Easy: A Practitioner's Guide to the RUP. Addison-Wesley, Boston (2003)

50. Vavpotic, D., Bajec, M.: An Approach for Concurrent Evaluation of Technical and Social Aspects of Software Development Methodologies. Information and Software Technology 51(2), 528–545 (2009)

ArchiMate and DEMO – Mates to Date?

Roland Ettema[1] and Jan L.G. Dietz[2]

[1] Logica Consulting Netherlands
`roland.ettema@gmail.com`
[2] Delft University of Technology, The Netherlands
`j.l.g.dietz@tudelft.nl`

Abstract. ArchiMate is an approach to modeling the architecture of enterprises. In the corresponding architecture framework, three enterprise layers are distinguished: business, application and technology. Although ArchiMate is broadly applied in practice, its semantics appears to be undefined. DEMO is a methodology for enterprise engineering that is facing a rapidly growing acceptance. It is firmly rooted in a sound and appropriate theoretical basis. DEMO also distinguishes between three enterprise layers: ontological, infological and datalogical. This paper reports on a theoretical and practical comparative evaluation of ArchiMate and DEMO. Only the business layer of ArchiMate and the ontological layer of DEMO are considered. Three conclusions are drawn. First, the two approaches are hardly comparable since ArchiMate belongs to the second and DEMO to the third wave of approaches. Second, the business layer of ArchiMate corresponds to all three layers of DEMO, without a possibility to distinguish between them. Third, ArchiMate could benefit from adopting DEMO as its front-end approach, thereby enforcing the rigorously defined semantics of DEMO on the Archimate models.

Keywords: ArchiMate, DEMO, Enterprise Engineering, Enterprise Architecture, Enterprise Ontology.

1 Introduction

In the ongoing turmoil of emerging paradigms, methods and techniques in the field of Enterprise Engineering, and the sometimes hot theoretical and practical discussions concerning them, occasionally approaches pop up that survive despite their actual or alleged shortcomings. Two current examples of such approaches, incidentally both originating from The Netherlands, are Archimate and DEMO. We have selected them for investigation and assessment because we think that a combination of the two could be beneficial. ArchiMate is a modeling language for enterprise architecture. The main information source regarding Archimate was [11]; however, since it has become a standard of The Open Group (TOG), the authoritative source is now the description

A. Albani, J. Barjis, and J.L.G. Dietz (Eds.): CIAO!/EOMAS 2009, LNBIP 34, pp. 172–186, 2009.

by TOG[1]. DEMO (Design and Engineering Methodology for Organizations) is an enterprise engineering methodology. The authoritative source for DEMO is [4]. This paper reports on the research we have conducted for the sake of articulating and elucidating the differences between Archimate and DEMO, in order to enable a sensible assessment of a possible combination. We have done that in the context of the new discipline of Enterprise Engineering. Although this discipline is certainly not fully established yet, the main characteristics are becoming clear [8]. They are summarized in the Enterprise Engineering Manifesto[2].

One of these characteristics is that Enterprise Engineering is the result of the merging of (the current state of) the Information Systems Sciences and the Organization Sciences. Within the former, three phases or waves can be distinguished in the understanding of the application of Information and Communication Technology (ICT) to enterprises. The first wave started with the introduction of computers in the sixties, and ended in the seventies of the 20th century. The few available approaches at that time focused on the form of information. Applying ICT basically meant replacing a paper document by its electronic equivalent. During the seventies a 'revolution' took place, pioneered by Langefors [10], who suggested to focus on the content of information before bothering about its form. Applying ICT began to mean automating information (i.e. content) needs, regardless the form in which the information is stored or presented. It marks the beginning of the second wave. Around 2000 another 'revolution' started, pioneered by people from the Language-Action Perspective community [6]. This marks the beginning of the third wave. Basing their insights on language philosophy [1, 7, 12], they suggested to recognize the intention of information on top of its content and to focus on this aspect first, before bothering about the content and the form. Examples of intentions are: request, promise, state, and accept. Because the informational notion of intention is closely related to the organizational notions commitment and responsibility, the 'natural' merge became possible of the information system sciences and the organizational sciences into the discipline of enterprise engineering. Since ArchiMate is based on the descriptive notion of architecture [5], we can safely equate the architecture of an enterprise with a conceptual model of its business processes and objects. From the description of Archimate it becomes clear that it is a second wave approach, meaning that it ignores the intention aspect of communication and information. DEMO is clearly a third wave approach. Yet, its scientific foundation is broader. Next to language philosophy, it includes system ontology [2] and world ontology [16].

Another main characteristic of Enterprise Engineering is a profound understanding of the process of system development of any kind, thus also of enterprises. For changing an enterprise, in particular for supporting its operational activities by means of ICT applications, one needs to have and appropriate understanding of the (stable) essence of an enterprise. From the engineering sciences in general it is known that if one wants to change a system, something of it must remain the same. For example, if

---

[1] www.opengroup.org/archimate
[2] See www.ciao-network.org

one wants to redesign a meeting room, it is important that it remains a meeting room. As another example, if one wants to support or even replace the employees in an accounting department by means of an automated accounting system, the accounting process must essentially stay untouched. So, in general one needs to have an understanding of a thing to be changed at a level just above the level at which the changes take place. If this understanding is lacking, one even cannot evaluate a change sensibly. For a correct understanding of the process of system development, DEMO relies on the Generic System Development Process [5]. Regarding Archimate, we did not find something similar. This reinforces the observation made earlier, that Archimate is only a modeling language, not a methodology.

Next to the theoretical investigation of Archimate and DEMO, we make a practical comparison by applying both to the case Car Import, that is taken from the ArchiMate project deliverable D3.5.1b [9]. Below the original narrative description is presented:

In The Netherlands any imported car is subjected to special kind of taxation called BPM. The business architecture supporting the whole collection process and the interaction of the Dutch Tax Department (Belastingdienst) with the importers and a number of other parties is described below. The importer (private person or car dealer/importer) must announce himself at the customer counter in any of the 30 Customs units in The Netherlands with the imported vehicle, its (provenance) documents, the approval proof of its technical inspection, and possibly with cash for the payment of the BPM tax. The public servant will handle the tax declaration as follows: first he will check all the documents, then he will fill in all the data into a client BPM application (running on a local server) and will calculate the due BPM tax value (using the BPM application and the catalogue value for that particular car). One copy of the BPM form (BPM17 ex 1) will be issued and sent to the administration. Another copy of this form is handed to the importer (BPM17 ex3), together with either the evidence of a cash payment (if the importer is able to pay the BPM amount in cash), or with a bill ("acceptgiro") issued for the due amount (in the case the importer is not able to pay in cash).

At each Customs unit there will be public servants assigned to handle the additional administrative operations regarding all the incoming BPM statements. Once a day, this person will collect all the incoming BPM17 forms. For ones, which were paid in cash, he will issue and authorize another copy of the BPM form (BPM17 ex2). This copy will be sent to RDW ("Rijksdienst voor het Wegverkeer" - the Netherlands Road Transport Department), which keeps the evidence of all registered vehicles in The Netherlands. The first copy of BPM 17 will be then sent to the archive. The forms which are not yet paid, are kept "on hold" until they are paid. The payment administration and the notification service for all incoming payments for these BPM forms is done by a separate department of the Belastingdienst, namely the Tax Collection Departments ("Inning"), which is responsible for the collection of all payments via bank. Once such a notification is received (via the BPM server application) the administration will prepare, authorize and send the copy of BPM17 ex.2 to RDW, and will permanently archive the ex1 of the BPM17.

The remainder of the paper is organized as follows. Section 2 summarizes the ArchiMate approach, as well as the analysis of the case Car Import with ArchiMate. Section 3 summarizes the DEMO approach, as well as the DEMO analysis of the case. In Section 4 we compare the two approaches, both theoretically and on the basis of the respective analyses of the case Car Import. Section 5 contains some salient conclusions regarding the differences as well as regarding the possible combination of ArchiMate and DEMO.

2 ArchiMate

2.1 Summary of ArchiMate

ArchiMate is a language for modeling enterprise architectures in accordance with a meta model and a conceptual framework of modeling concepts, called the ArchiMate Framework. ArchiMate is based on the descriptive notion of architecture [8], which means that an enterprise architecture in ArchiMate corresponds to a conceptual model of the business processes in the enterprise. The ArchiMate Framework is exhibited in Fig. 1. Three architectural layers are distinguished, called the business layer, the application layer, and the technology layer. The idea behind this division is that the application layer provides services to the business layer, and that the technology layer provides services to the application layer. Moreover, the business layer is said to provide business services to the environment of the enterprise.

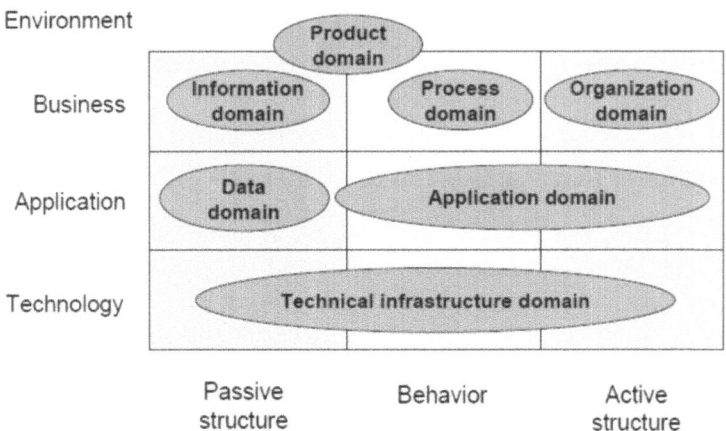

Fig. 1. The ArchiMate Framework

On the horizontal axis, three major aspects are distinguished, called active structure, behavior, and passive structure. The first two refer to generic system aspects, as e.g. identified by Weinberg [15]. The third aspect is related to the discipline of information system engineering. ArchiMate's perspective on organizations is highly influenced by this discipline. The meta model (see Fig. 2) is structured in conformity with the framework of Fig. 1.

The meta model therefore consists of active structural elements, of behavioral elements and of passive structural elements. Fig. 2 shows the meta model for the business layer. The concepts on the right hand side regard the active structure aspect. The concepts in the centre regard the behavioral or dynamic aspect, and the concepts on the left hand side regard the passive aspect.

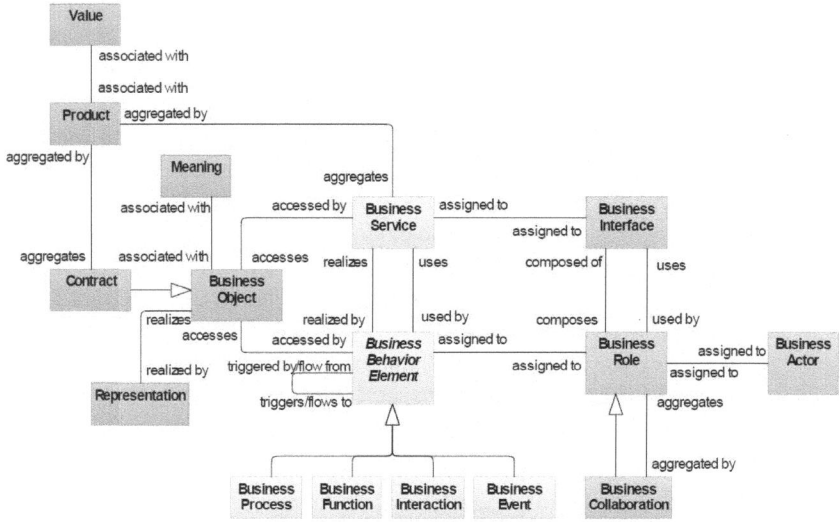

Fig. 2. ArchiMate Meta Model of the business layer

2.2 Analysis of the Case Car Import with ArchiMate

The narrative description of the case Car Import, as presented in section 1, constitutes the starting point for the modeling activity with ArchiMate. The first methodological step is to identify text elements that can be recognized as ArchiMate concepts. The second step is to position these elements within the framework and to determine the relationships that exist between them. The source from which we take the ArchiMate analysis of the case [9] does not provide further details about the modeling activity that has lead to the result as exhibited in Fig. 3. It merely only presents this result.

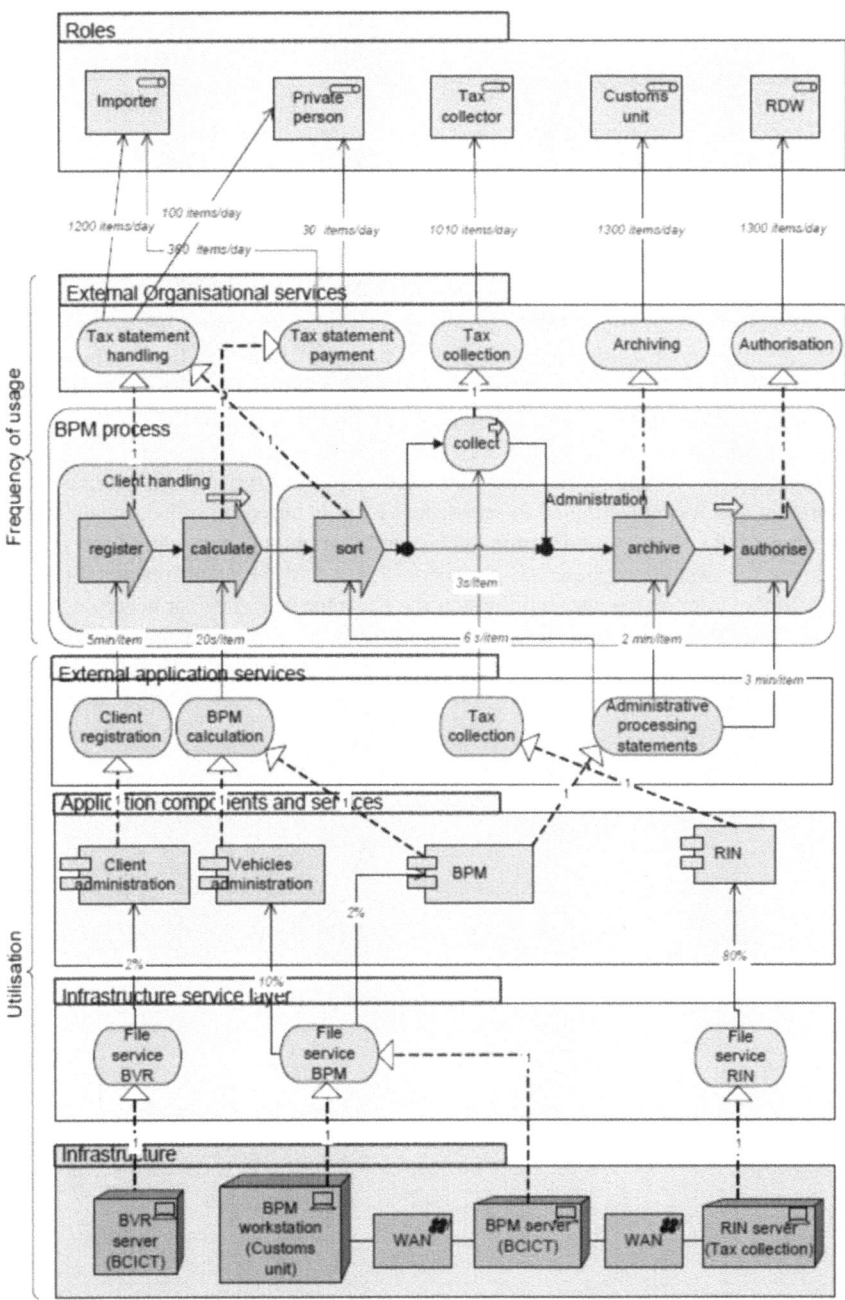

Fig. 3. ArchiMate model of the case Car Import

3 DEMO

3.1 Summary of DEMO

DEMO relies fully on the Ψ-theory [4]. In this theory, an enterprise (organization) is a system in the category of social systems [2]. The distinctive property of social systems is that the active elements are human beings or subjects. These subjects perform two kinds of acts: *production* acts (P-acts for short) and *coordination* acts (C-acts for short). By performing P-acts the subjects contribute to bringing about the goods or services that are delivered to the environment. By performing C-acts subjects enter into and comply with commitments towards each other regarding the performance of P-acts. Examples of C-acts are "request", "promise" and "decline". The effect of performing a C-act is that both the performer and the addressee of the act get involved in commitments regarding the bringing about of the corresponding P-act.

C-acts and P-acts appear to occur as steps in a generic coordination pattern, called *transaction*. Fig. 4 exhibits the basic transaction pattern (upper right corner), as the elaboration and formalization of the workflow loop as proposed in [3], which is drawn in the upper left corner. A transaction evolves in three phases: the order phase (O-phase for short), the execution phase (E-phase for short), and the result phase (R-phase for short). In the order phase, the initiator and the executor negotiate for achieving consensus about the P-fact that the executor is going to bring about. The main C-acts in the O-phase are the request and the promise. In the execution phase, the P-fact is brought about by the executor. In the result phase, the initiator and the executor negotiate for

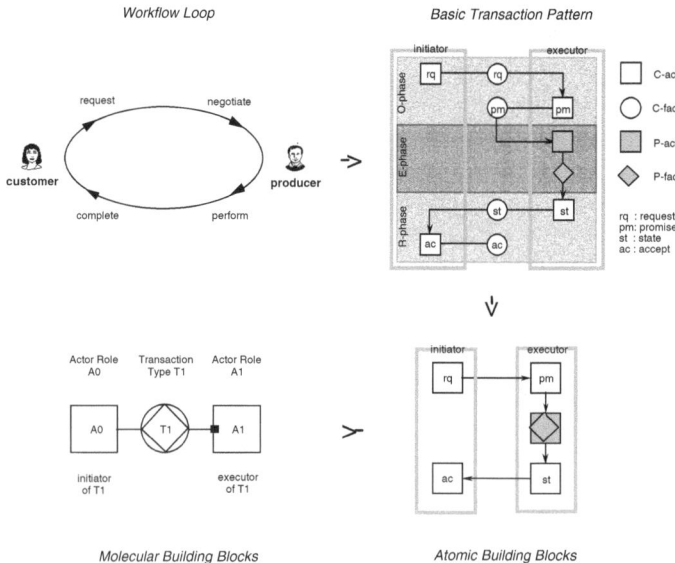

Fig. 4. Ontological building blocks of an organization

achieving consensus about the P-fact that is actually produced (which may differ from the requested one). The main C-acts in the R-phase are the state and the corresponding accept. The terms "initiator" and "executor" replace the more colloquial terms "customer" and "producer". Moreover, they refer to actor roles instead of subjects. An *actor role* is defined as the authority and responsibility to be the executor of a transaction type. Actor roles are fulfilled by subjects, such that an actor role may be fulfilled by several subjects and a subject may fulfill several actor roles.

The actual course of a transaction may be much more extensive than the basic pattern in Fig. 4. This is accommodated in the Ψ-theory by appropriate extensions of the basic pattern. At the lower right side of Fig. 4, a comprised notation is shown of the basic transaction pattern. A C-act and its resulting C-fact are represented by one, composite, symbol; the same holds for the P-act and the P-fact. At the lower left side the complete transaction pattern is represented by only one symbol, called the transaction symbol; it consists of a diamond (representing production) embedded in a disk (representing coordination). Transaction types and actor roles are the *molecular* building blocks of business processes and organizations, the transaction steps being the *atomic* building blocks.

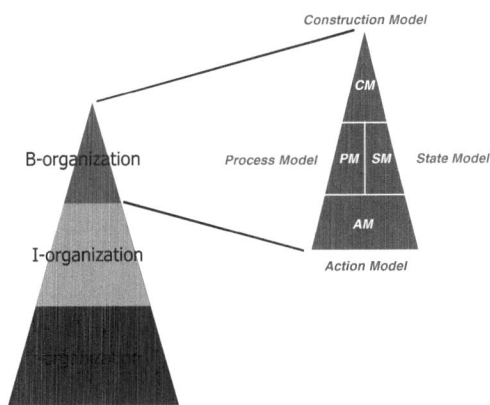

Fig. 5. The three aspect organizations

Another important component of the the Ψ-theory is the distinction between three human abilities, which are exerted both in C-acts and in P-acts: the forma, the informa, and the performa ability. Regarding coordination, the forma ability concerns uttering and perceiving written or spoken sentences, the informa ability concerns formulating thoughts and educing them from perceived sentences, and the performa ability concerns getting engaged in commitments. On the production side, the forma ability concerns datalogical production (storing, transmitting, copying etc. of data), the informa ability concerns infological production (computing, reasoning), and the performa ability concerns bringing about original new facts (deciding, judging, creating); we therefore call it ontological production.

The distinction between the three human capabilities on the production side gives rise to the distinction of three layered aspect organizations, as depicted in Fig. 5. By definition, the ontological model of an enterprise is the model (according to the Ψ-theory) of its B-organization. DEMO helps in 'discovering' an enterprise's ontological model, basically by re-engineering from its implementation, as e.g. contained in a narrative description. The complete ontological model of an enterprise consists of four aspect models (see Fig. 5). The Construction Model contains the actor roles and transaction kinds, the Process Model contains the business processes and business events, the State Model contains the business objects and business facts, and the Action Model contains the business rules.

3.2   Analysis of the Case Car Import with DEMO

Every experienced DEMO analyst has his or her own way of working in producing the DEMO models of a case, being fully guided by the Ψ-theory. For novice DEMO analysts, however, a six-step method has been developed [4]. Applying the first steps of this method to the narrative description of the case Car Import produces the result as presented hereafter.

In The Netherlands any [imported car] is subjected to special kind of taxation called BPM. The business architecture supporting the whole collection process and the interaction of the [Dutch Tax Department (Belastingdienst)] with the importers and a number of other parties is described below. The [importer] (private person or car dealer/importer) must announce himself at the customer counter in any of the 30 Customs units in The Netherlands with the <imported vehicle>, its (<provenance>) documents, the <approval proof of its technical inspection>, and possibly with cash for the payment of the BPM tax. The public servant will handle the tax declaration as follows: first he will check all the documents, then he will fill in all the data into a client BPM application (running on a local server) and will calculate the due BPM tax value (using the BPM application and the catalogue value for that particular car). One copy of the BPM form (BPM17 ex 1) will be issued and sent to the administration. Another copy of this form is handed to the importer (BPM17 ex3), together with either the (evidence of a cash payment) (if the importer is able to pay the BPM amount in cash), or with (a bill ("acceptgiro")) issued for the due amount (in the case the importer is not able to pay in cash).

At each Customs unit there will be [public servants] assigned to handle the additional administrative operations regarding all the incoming (BPM statements). Once a day, this person will collect all the incoming BPM17 forms. For ones, which were <paid> in cash, he will issue and <authorize> another copy of the BPM form (BPM17 ex2). This copy will be sent to RDW ("Rijksdienst voor het Wegverkeer" - the Netherlands Road Transport Department), which keeps the evidence of <all registered vehicles> in The Netherlands. The first copy of BPM 17 will be then sent to the archive. The forms which are not yet paid, are kept "on hold" until they are paid. The payment administration and the notification service for (all incoming payments) for these BPM forms is done by a separate department of the Belastingdienst, namely the Tax Collection Departments ("[Inning]"), which is responsible for the collection of all <payments> via bank. Once such (a notification is received) (via the BPM server application) the administration will prepare, <authorize> and send the copy of BPM17 ex.2 to [RDW], and will permanently archive the ex1 of the BPM17.

All ontological things are underlined. In addition, actors are indicated by placing their name between "[" and "]", P-acts and –facts are indicated by placing their name between "<" and ">", and C-acts and –facts are indicated by placing their name between "(" and ")". Next, we put the transaction pattern 'over' the onological things. This results in the identification of three transaction kinds: T01 - the import of a car, T03 - the admission of a car to the Dutch road network, and T04 - the payment of the BPM tax.

T01 is actually outside the scope of the case, but we will start to include it in our model since it clarifies the whole process from importing a car through to admitting it to the road network, and since paying BPM tax will turn out to be disconnected from importing a car, although the case description suggests so. T03 is only slightly mentioned, namely in the last sentence: ... *the administration will prepare, authorize and send the copy of BPM17 ex.2 to RDW* ... This sentence, inparticular the term "authorize" suggests that the sending of the copy counts as requesting for admission to the road network. However, this cannot be the case from an ontological point of view: only a car owner is authorized to request for admitting the car to the road network, as also only a car owner is authorized to request for importing the car.

Another, related, sentence that is ontologically puzzling, is the third one: *The importer (private person or dealer/importer) must announce himself at the customer counter in any of the Customs units* ... The question to be answered is "Who is requesting the importer to pay the BPM tax?". A candidate actor role is the one that decides on the import of a car. However, although the case description suggests that paying the BPM tax is connected to importing a car, this is not true, as further investigation has learnt us. The tax one has to pay as a prerequisite for importing a car is the VAT. We have included this transaction for completeness sake (T02). Importing a car however is distinct from getting it admitted to the road network! One could do the first and omit the second. So, there must be another actor role that requests to pay the BPM tax. Since paying this tax is a prerequisite for getting the car admitted to the road network, it is obvious (and institutionally quite correct) that RDW requests the car owner to pay the BPM tax after the car owner has requested the RDW to admit the car to the road network. Concludingly, we arrive at the Actor Transaction Diagram as exhibited in figure D1. The corresponding Transaction Result Table is shown in Table D1. Together they constitute the Construction Model (Fig. 5).

As said before, the left part of Fig. 6 was only included for the sake of explaining clearly the distinction between importing a car (including paying the VAT) and admitting a car to the road network (including paying the BPM tax). Fig. 6 clearly shows that the two processes are disconnected. Therefore, we only produce the Process Model for the right part, T03 and T04 (see Fig. 7). As a help in understanding it we have added to each step the person or organizational unit or institution that actually performs the step. For the sake of simplicity we have chosen CA03 and CA04 to be fulfilled by a private person. Obviously, RDW is the authorized institution for fulfilling actor role A02 (road network admitter). However, for performing T04/ac it has apparently delegated its authority to the Tax Office (Belastingdienst). The dashed arrow from T04/ac to T03/ex means that RDW has to wait for deciding to admit a car to the road network until the BPM tax has been paid. From the case description we derive that the current way in which RDW is informed about this fact by the Belastingdienst is the sending of the copy of BPM17 ex.2.

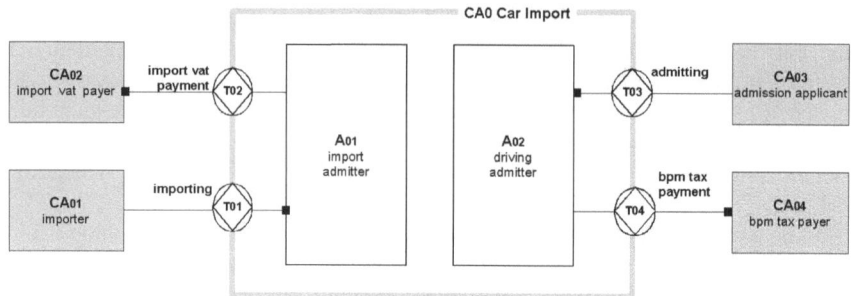

Fig. 6. Actor Transaction Diagram (ATD) of the case Car Import

Table 1. Transaction Result Table of the case Car Import

Transaction type	Transaction result
T01 importing	R01 *Import I has been performed*
T02 import vat payment	R02 *import vat for Import I has been paid*
T03 admitting	R03 *Admission A has been started*
T04 bpm tax payment	R04 *BPM Tax for admission A has been paid*

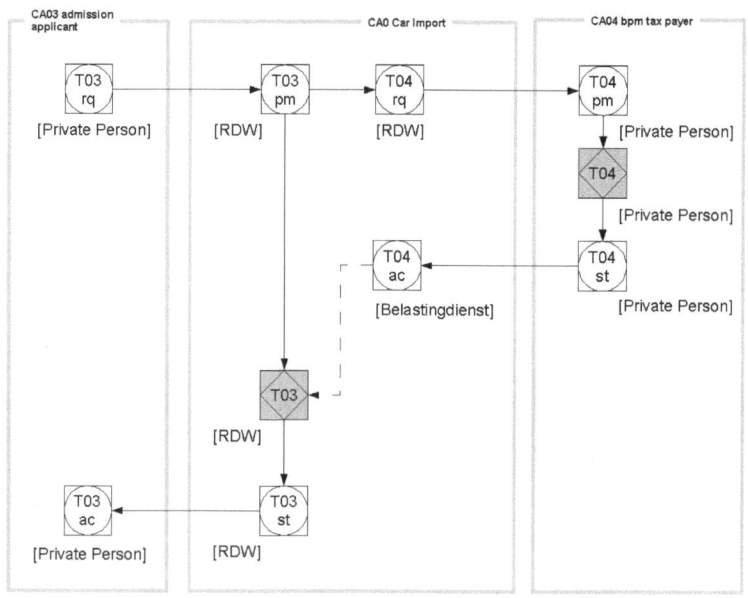

Fig. 7. Process Model (PM) of the case Car Import

4 Comparing ArchiMate and DEMO

4.1 Theoretical Comparison

By the theoretical comparison of ArchiMate and DEMO we mean the comparative evaluation of the Way of Thinking as well as the Way of Modeling of each, in accordance with the evaluation framework that is known as the 5-way model [13].

By the Way of Thinking of an approach is understood its theoretical foundation, in particular the basic understanding of the object of analysis, in our case the enterprise. At first sight, the business layer in ArchiMate seems to correspond with the B-organization in DEMO. This appears not to be true, however. To clarify the difference, we present in Fig. 8 the relationship between an organization and its supporting ICT-systems as conceived in DEMO.

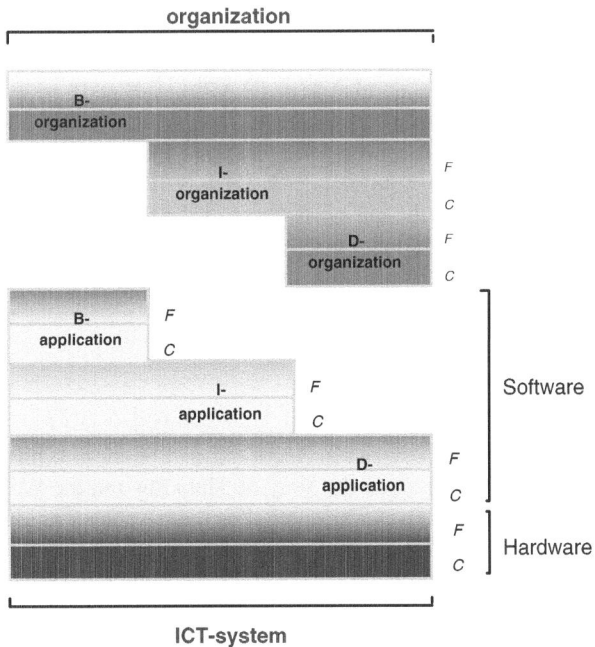

Fig. 8. Organization and ICT-system

Apparently, the business layer in ArchiMate corresponds to the three organization layers in DEMO (B-, I- and D-) collectively. Most probably, the application layer in ArchiMate corresponds to the B-application and the I-application layer in DEMO, and the technology layer in ArchiMate corresponds to the D-application and the hardware layer in DEMO. However, since we have focused on the business layer, there is no evidence to verify or falsify this hypothesis. Next, the Ψ-theory underlying DEMO provides for an appropriate and rigorous foundation. ArchiMate lacks such a foundation.

As a consequence, the semantics of the meta model (cf. Fig. 1) is undefined, which may easily lead to misinterpretations.

By the Way of Modeling (WoM) of an approach, is understood the definition of the distinct models of an enterprise, their representations, and the derivation of the models from a case description. The WoM of ArchiMate is to look for terms in the case description that designate instances of the meta model concepts. In this manner, the model as represented in Fig. 3 is produced. This WoM resembles very much the one in several modeling techniques from the seventies, of which the ER model is probably the best known. The advocated WoM was to look for nouns and verbs in a text. Nouns were taken as names of entity types and verbs ware taken as names of relationship types.

Despite the fact that ArchiMate's WoM (and meta model) is widely used, it has serious drawbacks. One is that irrelevant concepts are included in the model, just because there was a term in the description referring to it. Another one is that relevant concepts are missing because references in the description were forgotten. A third one is that different analysts produce different models, since the meta model is multi-interpretable.

In contrast, since the four aspect models of DEMO (cf. Fig. 5) are grounded in the Ψ-theory, the ontological model of an enterprise is guaranteed coherent, consistent, comprehensive, and concise. Moreover, it only shows the essence of the enterprise, completely independent of any implementation issue. This not only holds for the B-organization (the essence of the enterprise) but also for the I-organization and the D-organization. Therefore, it the ideal starting point for the re- design and re- engineering of enterprises [5]. Lastly, different analysts will produce the same ontological model, because every enterprise has only one such model.

4.2 Comparing the Analysis Results

As one will have observed, the results of applying ArchiMate and DEMO to the case Car Import, as presented and discussed in Section 2 and Section 3, differ very much. This is due to the differences between the Way of Thinking and the Way of Modeling of the two approaches. Obviously, ArchiMate takes the case description literally; it is its Way of Modeling. On the other hand, the DEMO analysis has evoked 'critical' questions. They have lead to a profound understanding of what is essentially going on in the described situation.

First, importing a car and getting a car admitted to the Dutch road network are distinct and disconnected processes. Only for the latter one, there is the prerequisite that the BPM tax is paid.

Second, it is not true that the Tax Office authorizes the RDW to admit a car to the road network; instead it is the car owner who requests for admission. The Tax Office only has a delegated authority in accepting the results of transactions T04 (BPM tax payment). Subsequently it informs the RDW that the payment has been received. We have a strong suspicion that the Tax Office and the RDW are not aware of these essential relationships. It is for sure, however, that this ignorance causes a lot of confusion and many failures in attempts to make the processes more efficient, e.g. by applying modern ICT.

5 Conclusions

We have carried out a comparative evaluation of ArchiMate and DEMO, both theo-retically and practically, i.e. on the basis of the analysis of the same case by each approach. Space limitations prohibit us from giving a full and detailed account of our research. Only the most noticeable issues could be presented and discussed. In addi-tion, a thorough assessment of the strengths and weaknesses of ArchiMate and DEMO can only be performed on the basis of multiple real-life and real-size cases, taken from different areas. Nevertheless, some conclusions can certainly be justified already now.

The first conclusion is that ArchiMate and DEMO are hardly comparable, for sev-eral reasons. One is that ArchiMate is a second wave approach, whereas DEMO is a third wave approach, as was discussed in Section 1 already. Another reason is that DEMO is founded in a rigorous and appropriate theory, whereas ArchiMate lacks such a foundation. Therefore, its semantics are basically undefined, which unavoid-ably leads to miscommunication among Archimate users. One would expect that having a rigorous semantic definition would be a prerequisite for an open standard.

A second conclusion regards the abstraction layers as disinguished by ArchiMate and DEMO. DEMO (in fact the Ψ-theory) makes a well defined distinction between three abstraction layers: the B-organization, the I-organization, and the D-organization. Only in the B-organization original new production facts are brought about (deciding, judging, manufacturing etc.), by which the enterprise world is changed. In the I-organization one computes, calculates, reasons; this does change the world. In the D-organization one stores, copies, transports etc., documents. Despite the fact that ArchiMate belongs to the second wave, it does not make a distinction between in-fological and datalogical issues in the business layer. As an illustration of the point, the model in Fig. 3 includes actions like archiving and sorting, next to calculation. Al-though this seems not to be an issue of worry for Archimate, we think ArchiMate could profit from solidly incorporating this distinction. It would make Archimate to some extent suitable for re-engineering projects. The lack of a rigorous semantic definition remains a major obstacle for actually doing it.

Although ArchiMate and DEMO are to a large extent incomparable, we think that they can usefully be combined. As a matter of fact, several studies have been carried out concerning the combination of DEMO with some second generation approach, since DEMO does not really cover the implementation of an organization. An inter-esting study in this respect is an evaluative comparison of DEMO and ARIS, in par-ticular the EPC (Event Process Chain) technique [14]. As one of the practical outcomes, a procedure has been developed for producing EPCs on the basis of DEMO models. In this way, the rigorous semantics of DEMO are so to speak enforced upon the EPC. We conjecture that such a combination is also possible and beneficial for ArchiMate and DEMO.

References

1. Austin, J.L.: How to do things with words. Harvard University Press, Cambridge (1962)
2. Bunge, M.A.: Treatise on Basic Philosophy. A World of Systems, vol. 4. D. Reidel Pub-lishing Company, Dordrecht (1979)

3. Denning, P., Medina-Mora, R.: Completing the loops. In: ORSA/TIMS Interfaces, vol. 25, May 3-June, pp. 42–57 (1995)
4. Dietz, J.L.G.: Enterprise Ontology – theory and methodology. Springer, Heidelberg (2006)
5. Dietz, J.L.G.: Architecture – building strategy into design, Sdu Netherlands (2008)
6. Special issue of Communications of the ACM 49(5), 59–64 (May 2006)
7. Habermas, J.: Theorie des Kommunikatives Handelns, Erster Band. Suhrkamp Verlag, Frankfurt am Main (1981)
8. Hoogervorst, J.A.P., Dietz, J.L.G.: Enterprise Architecture in Enterprise Engineering. Enterprise Modelling and Information Systems Architecture 3(1) (March 2008)
9. Iacob, M.-E., Jonkers, H.: Quantitative Analysis of Enterprise Architectures. Enschede: Telematica Instituut, Archimate Deliverable 3.5.1b/v2.0. TI/RS/2004/006 (2004)
10. Langefors, B.: Information System Theory. Information Systems 2, 207–219 (1977)
11. Lankhorst, M., et al.: Enterprise Architecture at Work. Springer, Heidelberg (2005)
12. Searle, J.R.: Speech Acts, an Essay in the Philosophy of Language. Cambridge University Press, Cambridge (1969)
13. Seligman, P.S., Wijers, G.M., Sol, H.G.: Analyzing the structure of I.S. methodologies; an alternative approach. In: Maes, R. (ed.) Proceedings of the First Dutch Conference on Information Systems, Amersfoort (1989)
14. Strijdhaftig, D.: DEMO and ARIS – developing a consistent coupling, Master Thesis TU Delft (2008)
15. Weinberg, G.M.: An Introduction to General Systems Thinking. John Wiley & Sons, Chichester (1975)
16. Wittgenstein, L.: Tractatus logico-philosophicus. Routledge & Kegan Paul Ltd., London (1922) (German text with an English translation by C.K. Ogden)

Integration Aspects between the B/I/D Organizations of the Enterprise

Joop de Jong

Delft University of Technology, Netherlands
Xprise Business Solutions, P.O. Box 598, 3900 AN Veenendaal, Netherlands
jdjong@xprise.com

Abstract. The enterprise can be considered as a heterogeneous system consisting of three homogeneous systems, called the B(usiness)-organization, the I(nformation)-organization and the D(ata)-organization. The D-organization supports the I-organization and the I-organization supports the B-organization. Those three organizations are linked to each other through the cohesive unification of the human being. However, from the construction point of view it is unclear how the interactions between actors of the different layers in the enterprise are taken place. This paper contributes to more clearness about this interface by elaborating some integration aspects between the mentioned layers more in detail.

Keywords: enterprise engineering, enterprise ontology, infological system, DEMO.

1 Introduction

Mulder gives in his PhD thesis an extended overview of the methodologies which have been developed by the organization sciences. His conclusion is that these sciences are not fully capable to develop a methodology in which the various aspects of organization design, as structure, processes, information systems, are taken into account in a integrated way [1]. His conclusion is that the key to the desired integration between all those aspects is to make the notion of communication as the central notion for understanding organizations. He refers to the relatively young research field of Language Action Perspective, or LAP for short. The focus on communication as the concept for understanding and modeling of organizations comes from the Speech Act Theory. This theory does not only consider speech acts as a vehicle to transfer knowledge, but also as a vehicle to act, by which new facts can be created [2, 3]. Based on this theory several methodologies have been developed. Mulder has compared the most important LAP-based methodologies in his thesis and comes to the conclusion that the Design and Engineering Methodology for Organizations DEMO is the most appropriate methodology to (re)design organizations in an integrated way. The major difference between DEMO and other LAP approaches is that it builds on two additional theoretical pillars next to the LAP, namely Organizational Semiotics [4, 5] and Systems Ontology [6]. The way of thinking and the way of modeling of

A. Albani, J. Barjis, and J.L.G. Dietz (Eds.): CIAO!/EOMAS 2009, LNBIP 34, pp. 187–200, 2009.

DEMO are developed by Dietz [7]. However, Dietz focuses primarily on the business layer of the enterprise and gives only some clues concerning the infological layer and the datalogical layer of the enterprise. The distinction between the three layers discerns DEMO from many other aggregated modeling techniques such as Petri Net, Flow Chart and EPC. In particular, a split between models of the three mentioned layers makes designing and engineering of enterprises in a dynamic world more intellectually in control. All layers have to be considered as aspect systems of the enterprise system. They are called the B-organization, the I-organization and the D-organization. Actors in the I-organization provide information services to actors in the B-organization which need this information for performing business transactions. The required data has been delivered by the D-organization. This paper presents the way on which the interactions between actors of the different layers take place.

Section 2 contains a summary to the Ψ-theory on which DEMO has been grounded. On the basis of the organization theorem the elementary components within the three layers and their mutual relationships are discussed in section 3. The paper is ended by section 4 with some conclusions and some directions for further research.

2 Summary of the Ψ-Theory

For a good understanding of this paper a summary of the Ψ-theory on which Dietz [7] based the DEMO methodology is presented. Dietz argues that in order to cope with the current and future challenges, a conceptual model of the enterprise is needed that is coherent, comprehensive, consistent and concise, and that only shows the essence of the operation of an enterprise model. Such a model, called an ontological model, abstracts from all implementation and realization issues. The underlying theory is called the Ψ-theory. The Ψ-theory consists of four axioms, viz. the operation axiom, the transaction axiom, the composition axiom and the distinction axiom, and the organization theorem. In this section, these axioms and the organization theorem are elaborated briefly. An exception is made for the composition axiom which is of no importance for this paper.

The *operation axiom* states that the operation of the enterprise is constituted by the activities of actors, which are elementary chunks of authority and responsibility fulfilled by human beings. Actors perform two kinds of acts: production acts, or P-acts for short, and coordination acts, or C-acts for short. These acts have definite results, namely production facts and coordination facts, respectively. By performing P-acts, actors contribute to bringing about the goods or services that are delivered to each other or to the environment. A P-act is either material or immaterial. Examples of material acts are manufacturing acts and storage and transportation acts. Examples of immaterial acts are the judgment by a court to condemn someone, granting an insurance claim and selling goods. By performing C-acts, actors enter into and comply with commitments towards each other regarding the performance of P-acts. A C-act is defined by its proposition and its intention. The proposition consists of a P-fact, e.g. "Purchase order #200 is delivered" and a time period (the delivery time). The intention represents the purpose of the performer; examples of intentions are "request", "promise" and "decline". The effect of performing a C-act is that both the performer and the addressee of the act get involved in a commitment regarding the referred P-act.

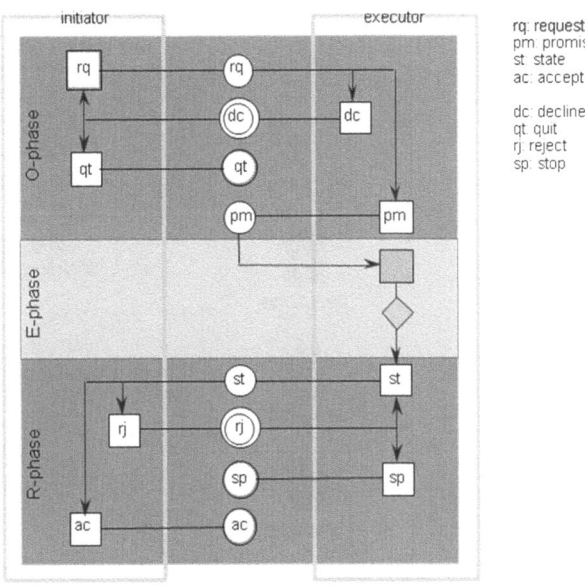

rq: request
pm: promise
st: state
ac: accept

dc: decline
qt: quit
rj: reject
sp: stop

Fig. 1. The standard transaction pattern

The *transaction axiom* states that coordination acts are performed as steps in universal patterns. These patterns, also called transactions, always involve two actor roles, i.e. two chunks of authority and responsibility. They are aimed at achieving a particular result, the P-fact.

Figure 1 exhibits the standard transaction pattern. A transaction evolves in three phases: the order phase (O-phase for short), the execution phase (E-phase for short) and the result phase (R-phase for short). One of the two partaking actor roles is called the initiator, the other the executor of the transaction. In the order phase, the initiator and the executor pursue to reach agreement about the P-fact that the executor is going to bring about as well as the intended time of creation. In the execution phase, the executor brings about this P-fact. In the result phase, the initiator and the executor pursue to reach agreement about the P-fact that is actually produced as well as the actual time of creation (both of which may differ from the requested one). Only if this agreement is reached will the P-fact become existent. The path request-promise-execute-state-accept in figure 1 is called the basic pattern; it is the course that is taken when the initiator and the executor keep consenting. However, they may also dissent. There are two states where this may happen, namely the states "requested" and "stated". Instead of promising one may respond to a request by declining it, and instead of accepting one may respond to a statement by rejecting it. It brings the process in the state "declined" or "rejected" respectively. These states are indicated by a double disk, meaning that they are discussion states. If a transaction ends up in a discussion state, the two actors must 'sit together', discuss the situation at hand and negotiate about how to get out of it. The possible outcomes are a renewed request or statement (probably with a modified proposition) or a failure (quit or stop).

The *distinction axiom* states that there are three distinct human abilities playing a role in the operation of actors, called performa, informa and forma (cf. fig. 2). Those abilities are recognized in both kinds of acts that actors perform.

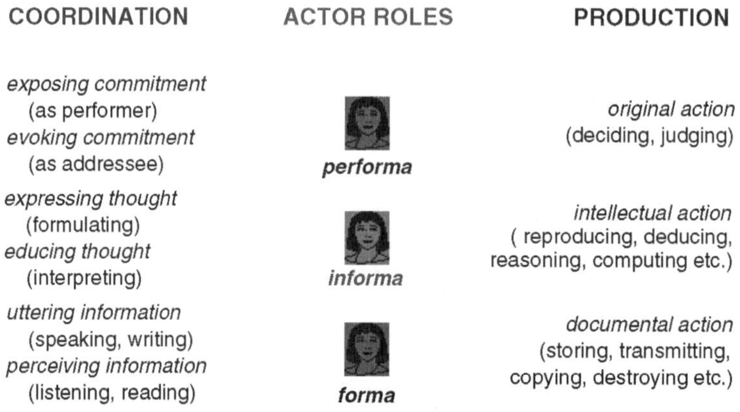

COORDINATION	ACTOR ROLES	PRODUCTION
exposing commitment (as performer) *evoking commitment* (as addressee)	**performa**	*original action* (deciding, judging)
expressing thought (formulating) *educing thought* (interpreting)	**informa**	*intellectual action* (reproducing, deducing, reasoning, computing etc.)
uttering information (speaking, writing) *perceiving information* (listening, reading)	**forma**	*documental action* (storing, transmitting, copying, destroying etc.)

Fig. 2. The three human capabilities

Let us first look at the production act of an actor. The forma ability is the human ability to conduct documental actions, such as storing, retrieving, transmitting, etc. These are all actions by which the content of the documents or data is of no importance. Actors which use the forma ability to perform P-acts are called documental actors, or D-actors for short. The informa ability is the human ability to conduct intellectual actions, such as reasoning, computing, remembering and recalling of knowledge, etc. These are all actions by which the content of data or documents, abstracted from the form aspects, is of importance. Actors which use the informa ability to perform P-acts are called intellectual actors, or I-actors for short. The performa ability is the human ability to conduct new, original actions, such as decisions, judgments etc. The performa ability is considered as the essential human ability for doing business, of any kind. It adds the notion of ontology on top of infology. Actors which use the performa ability to perform P-acts are called business actors, or B-actors for short. Human beings in enterprises are able to fulfill B-actors roles, as well as I-actor roles and D-actor roles.

Subsequently, let us look at the coordination act of an actor. By the notion of actor is meant a B-actor, as well as an I-actor or a D-actor. By performing C-acts, actors enter into and comply with commitments towards each other with respect to the performance of P-acts. The effect of performing a C-act is that both the performer and the addressee of the act get involved in a commitment concerning the referred P-act. That commitment is a result of a performative exchange. However the only way of the performer of the C-act to expose its commitment and to make it knowable to the addressee, is to express its informa ability, followed by the inducement in the mind of the addressee of an equivalent thought, by means of its informa ability. The intellectual understanding between both actors comes into existence by informative exchange. Expressing a thought can only be done by formulating it in a sentence in some language, and at the same time uttering it in some form, such as speaking or writing. Significational understanding between both actors only come into existence by a formative exchange, which means that the performer has to use a language that is known to the addressee. We would put with great emphasis that B-actors, I-actors and D-actors only distinguish themselves by the kind of production act.

The last part of the Ψ-theory is the *organization theorem*. It states that the organization of an enterprise is a heterogeneous system that is constituted as the layered integration of three homogeneous systems: the B-organization, the I-organization, and the D-organization. The D-organization supports the I-organization, and the I-organization supports the B-organization (cf. fig. 3). A system can be considered from two different perspectives, namely from the function perspective or from the construction perspective. The function perspective on a system is based on the teleological system notion which is concerned with the (external) behavior or performance of a system. This notion is adequate for the purpose of controlling or using a system. It is about services or products which are delivered by the system. The construction perspective on a system is based on the ontological system notion. Systems have to be designed and constructed. During the design and engineering process questions of being effectively and efficiency of the chosen design have to be answered by the constructor of the system. Our point of departure in this paper is the second system notion.

The D-organization supports the I-organization, and the I-organization supports the B-organization (cf. fig. 3). The integration is established through the cohesive unification of the human being. Let us elaborate this point more in detail. We take the I-organization as our starting point. From the functional perspective the I-organization provides an information service to the B-organization, i.e. to a B-actor which actually interprets the received data as required information in order to execute an ontological action, i.e. a C-act or a P-act. The information is produced by I-actors and is based on the C-facts and P-facts which have been created by B-actors from the B-organization or by external actors if the facts are retrieved from external fact banks. For example, the I-organization produces a monthly report with the turnover per product group. Actors within the I-organization have the competences and the authorities to construct such a report by making use of the available facts. However, they do not have any idea about the added value of this report for the salesman who has asked for it. I-actors are only producers of information for each other or for initiating business actors. However, how can a B-actor receive some information from a I-actor? According to the system definition of Bunge [6], the initiator of a transaction must be from the same category as the executor of the transaction. In other words, only an I-actor is allowed to request for information. The answer is given by the distinction axiom. By the cohesive unification of human being the needed I-actor could be a B-actor that has been shaped into an I-actor [7, 8].

For conducting an infological action an I-actor often has to reproduce existent facts. In order to reproduce such a fact the corresponding fact data must be retrieved from one of the fact banks within the system or from the environment of the system. Therefore, the mentioned I-actor shapes into a D-actor in order to initiate a D-actor for retrieving the needed fact data. Besides that, as a consequence of a particular infological action, an I-actor also have to remember new facts. That means actually that the corresponding fact data have to be recorded in a fact bank within the system in order to be used later on by an I-actor for a specific infological action. Recording the

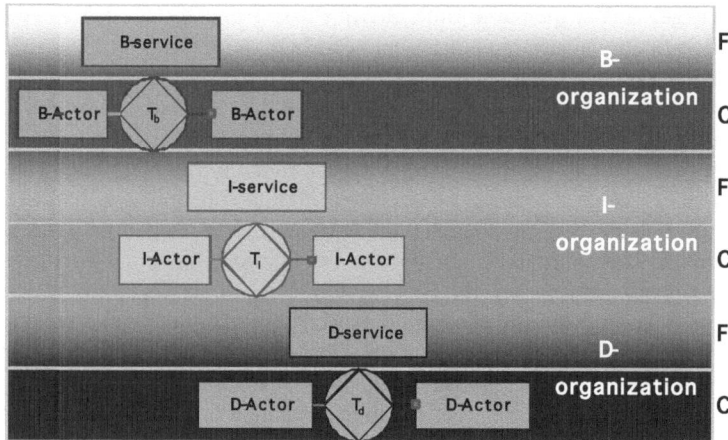

Fig. 3. The layered integration of an enterprise system

corresponding fact data in a fact bank is done by a D-actor which is initiated by the D-actor that has been shaped from the original I-actor.

Dietz [7, 9, 10] has focused mainly on the way of thinking and on the way of modeling with DEMO. Mulder [1] in his doctoral thesis presented a rapid enterprise design, a way of working with DEMO. All these publications have in common that they focus on the B-organization. The distinguishing feature of this paper is that the focus is moved from the B-organization to the I-organization and the D-organization. We discuss about differences between the B-actors, I-actors and D-actors and their mutual relationships within the organization boundary and their relationships with actors of the same category outside the organization boundary. The elaboration of these notions leads to a better understanding of the ontological models of the I-organization and the D-organization.

3   Cooperation in the Layered Enterprise

A subject who fulfils B-actor roles disposes of all three human abilities, as stated above. The subject uses the performa ability if it executes ontological actions. However, it is not hard to switch between abilities. If the subject takes the shape of an I-actor it is able to participate in I-transactions. After completing its infological transaction it switches back to its B-actor shape to resume its work in that shape [7, 8]. A particular subject can fulfill several B-actor roles, I-actor roles and D-actor roles within the same enterprise.

The raw materials for infological actions are C/P-facts which are defined by the B-actors as results of their coordination and production acts. These acts have effects on the coordination world or C-world and on de production world or P-world respectively [7]. A state of the P-world is a set of P-facts that have been created up to that point in time and the state of the C-world is a set of C-facts that have been created up to that point of time. So, we keep track on the complete history of both worlds. All

C-facts and P-facts are stored in coordination banks and production banks respectively. There are no other sources that contain raw materials for information production. However, not all relevant coordination and production banks are available within the organization boundary that one has defined for a particular enterprise. Those banks which are situated outside the organization boundary, called external coordination and production banks, can be consulted to obtain relevant C/P-facts as well. These C/P-facts are not been defined by B-actors of the B-organization of the mentioned enterprise but by B-actors of another enterprise working in the same P-world. For this paper it is of no importance to discern both mentioned types of fact banks. Hereafter, we only like to talk about fact banks. Although we do not discuss about the physical implementation of a fact bank, it is important to be aware that a fact bank can be embodied in a physical substrate in several ways, a database management system, a particular place in human brains, a measuring device, etc.

A B-actor shapes into an I-actor in order to initiate an infological transaction for two reasons. Firstly, it initiates a request for information which is based on existent facts that have to be reproduced. Secondly, it formulates a C/P-fact that has to be remembered. Both situations are discussed in this section. Let us start with the first one: the I-actor requesting for information.

Again two different situations could taken place, namely the needed information can be received from the I-organization, because of some processing that has to be done by the I-organization, or the information can be received directly from the D-organization by interpreting C/P facts which are offered to the D-actor which is the shaped initiating I-actor. To make the second situation more clear: if an I-actor requests for a C/P-fact then it receives the corresponding fact data as a D-actor and it interprets this fact data as an I-actor. There are no infological transactions required for getting the needed information. Infological transactions only occur if the requested information must be prepared from the infological perspective before delivery. The relationship between B-actors, I-actors and D-actors during information delivery is discussed further on the basis of an example that has been exhibited in figure 4 and 5.

Let us discuss first the example that the information is delivered by the I-organization (cf. fig. 4). We assume that John wants to perform an original action and that he needs some information to execute the original action in a correct way. According to the organization theorem John has to shape into a I-actor before he is able to request the required information. From that moment John fulfills an I-actor role within the I-organization. John requests Bill to send him the information he needs. If Bill is able to perform the infological action completely he will sent the results of this actions to John directly. If he needs other I-actors for producing parts of the requested information he has to wait for these subassemblies before he can satisfy John. In this example we assume that Bill performs the infological action entirely. Bill is not able to provide an information product without using existing facts. These facts have to be reproduced. Bill shapes into an D-actor and requests the D-actor Tom for a documental production result, namely the retrieval and transmission of the data that corresponds with the requested facts. Such an action can occur several times during the execution of an infological transaction. Tom performs, possibly in conjunction with other D-actors the demanded documental transactions.

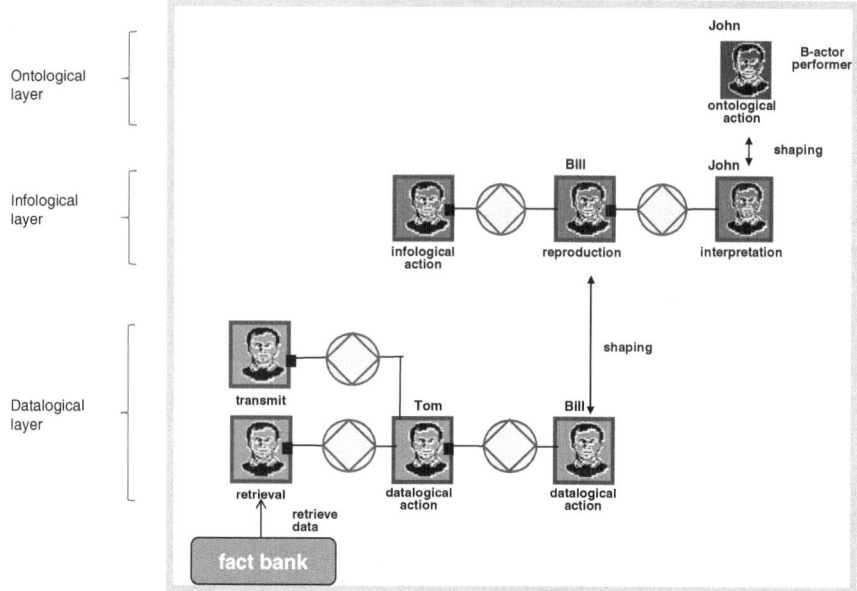

Fig. 4. Asking for information

On the basis of a practical example we can clarify this way of thinking more thoroughly. The example regards a validity check of a credit card used at a payment. The B-actor Peter would like to pay the B-actor John with his credit card. John formulates a request for Bill to check the validity of the credit card. He sends him the number and the expiring date of the card. Bill promises John an answer on his request. However, before he activates the checking process he has to come into the possession of some relevant data from Tom. After the receipt of this data Bill executes the process and provides John the outcome.

We mentioned in the previous section that a DEMO based model only shows the essence of the operation of an enterprise. Such a model, which is called an ontological model, abstracts from specific implementation issues. It is important to take this quality of the model into account if an example from practice is discussed on the basis of figure 4. Take in mind the following example about a salesman who would like to consult a product catalogue with detailed product information, prices and so on, during a sales transaction. From the view of ontology the phenomenon product catalogue is of no importance. It is the consequence of a particular implementation choice. The product information could also be stored on a compact disc or on the hard disc of the PC of the salesman or on anything else. More important is that product information is available for the salesman during a sales transaction. The ultimate sources of this product information are P-facts which are retrieved from fact banks. These P-facts are the result of original actions conducted by B-actors which have been developed the product information. Looking at figure 4, the salesman John asks Bill for the right product information. Bill promises that he will deliver this information and asks Tom for the fact data that corresponds with the concerning P-facts. Bill puts together all information received from Tom and offers it to John subsequently.

Figure 5 exhibits the situation that John requests for a C/P-fact which has been stored in a fact bank. For example John as actor in the enterprise operates autonomously. He constantly loops through the actor cycle, in which he deals with his agenda. An agendum is a C-fact with a proposed time for dealing with it , to which the actor is committed to respond. John shapes into an I-actor in order to reproduce the needed C-fact. Reproducing can be done by shaping into a D-actor and asking Tom for the corresponding fact data. Tom retrieves and transmits the found fact data subsequently to John for interpretation.

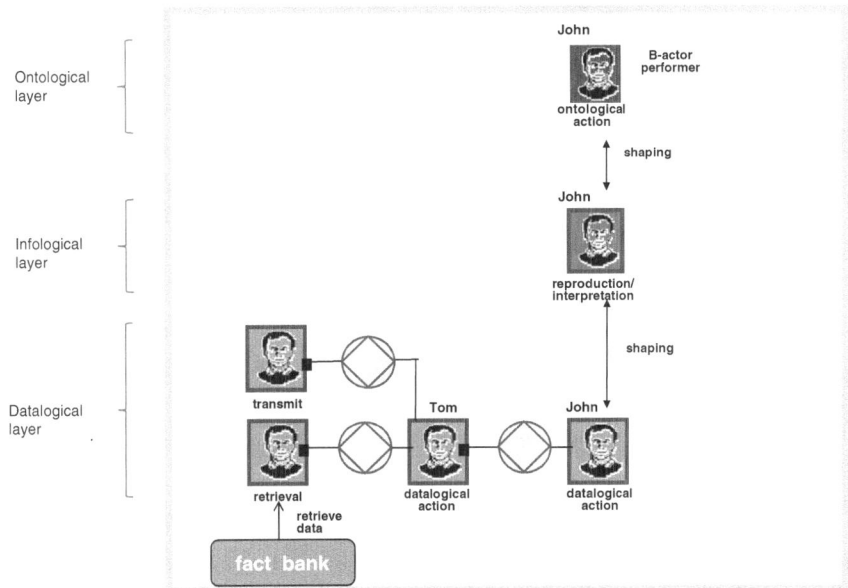

Fig. 5. Asking for C/P-facts

Next, let us discuss the second situation we mentioned: the I-actor remembering C/P-facts. A B-actor does not only shape into an I-actor for initiating information requests and interpreting received information but for formulating C/P facts and initiating requests for storing facts as well. Again, there are two different situations to discern. Firstly, a subject that shapes from a B-actor into an I-actor formulates a C/P-fact. Next, the subject shapes from an I-actor into a D-actor and offered the data that corresponds with the C/P-fact to the D-organization for storage. Secondly, this situation differs from the previous one that the C/P-fact is offered to the I-organization by the I-actor which has been shaped from the B-actor for remembering The figures 6 and 7 exhibit both situations respectively. Let us first look at figure 6. The B-actor John wants to store a C-fact or a P-fact. John shapes from a B-actor into an I-actor and formulates the C/P-fact. After that, he shapes from an I-actor into a D-actor and requests Tom to transmit and to record that particular fact into a fact bank. Therefore,

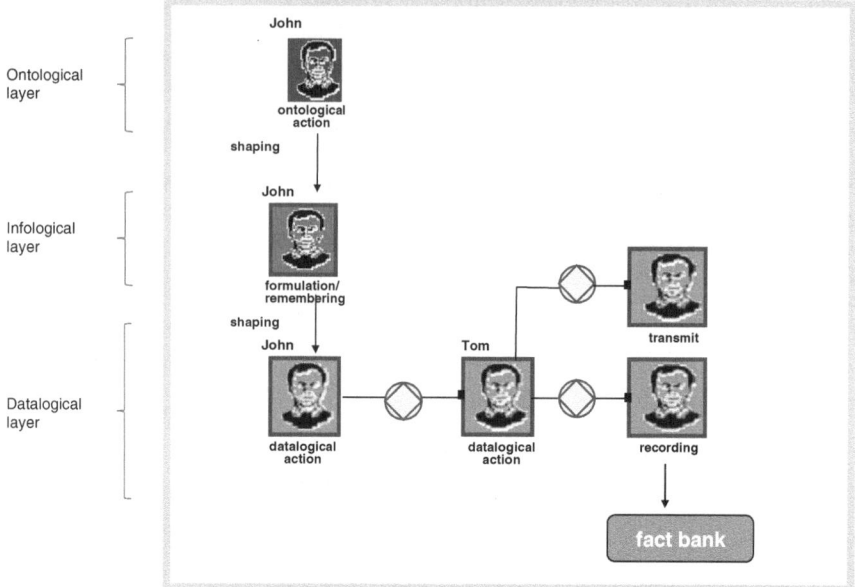

Fig. 6. Storing facts (1)

Tom closes transactions with some other D-actors in order to keep his promise. It strikes that no infological transaction is needed. It is comparable with the situation that John requests Tom to put some notes, which he has written down, into a book-case. No other I-actors are involved in formulating the notes. Figure 7 exhibits the situation that a C/P-fact is offered to the I-organization for remembering. John formulates the C/P-fact and requests Bill for remembering the formulated fact. Bill promises to remember the fact, shapes into a D-actor and offers the corresponding fact data to Tom for recording in a fact bank.

The mentioned four examples (cf. fig 4, 5, 6, 7) illustrate the possible interactions between the different layers in the organization. The I-organization supports the B-organization by the delivery of information services. These information services are developed by I-actors based on C/P-facts which are stored inside or outside the organization boundary and are reproduced by them. The C/P-facts inside the organization boundary are defined by B-actors within the B-organization as a result of their coordination and production acts. Those C/P-facts are offered to the I-organization for remembering. The D-organization supports the I-organization by the delivery of data services for reproducing and remembering C/P-facts.

The figures 4 till 7 contain a grey frame which indicates the organization boundary. An organization boundary divides the set of all (relevant) actor roles into the composition and the environment [7]. In those examples, the cooperation between the actors and the retrieval and record of data is taken place within the same organization. However, an interesting question arises about what will happen if a B-actor needs information which is only obtainable outside the organization, i.e. in external fact banks or

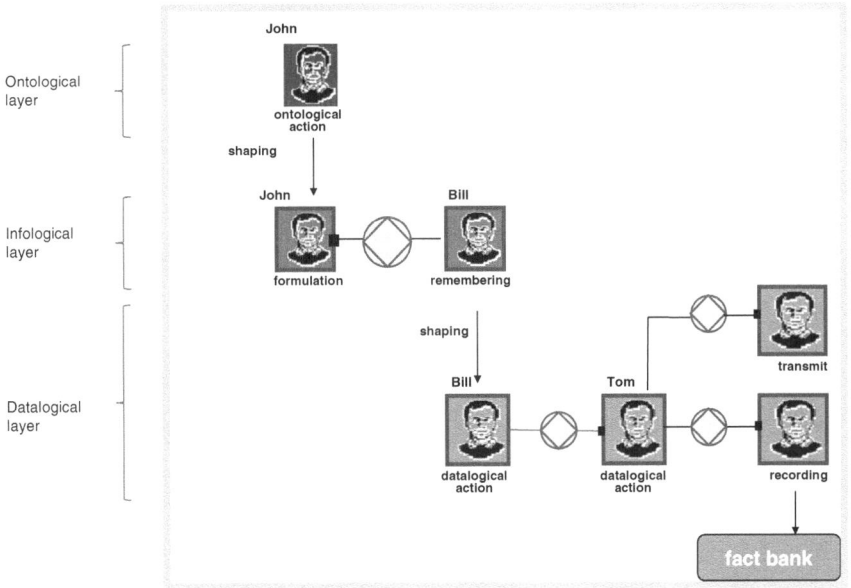

Fig. 7. Storing facts (2)

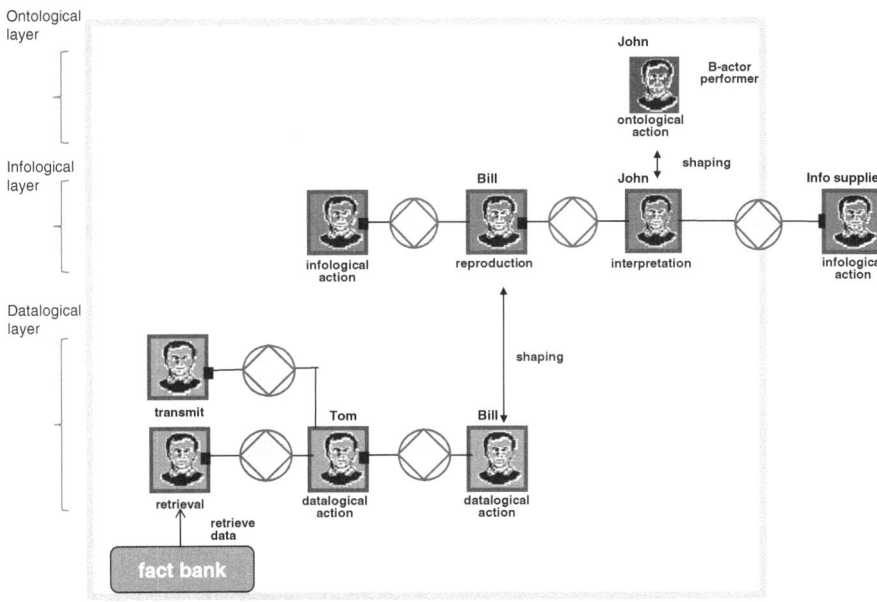

Fig. 8. Asking for internal and external facts

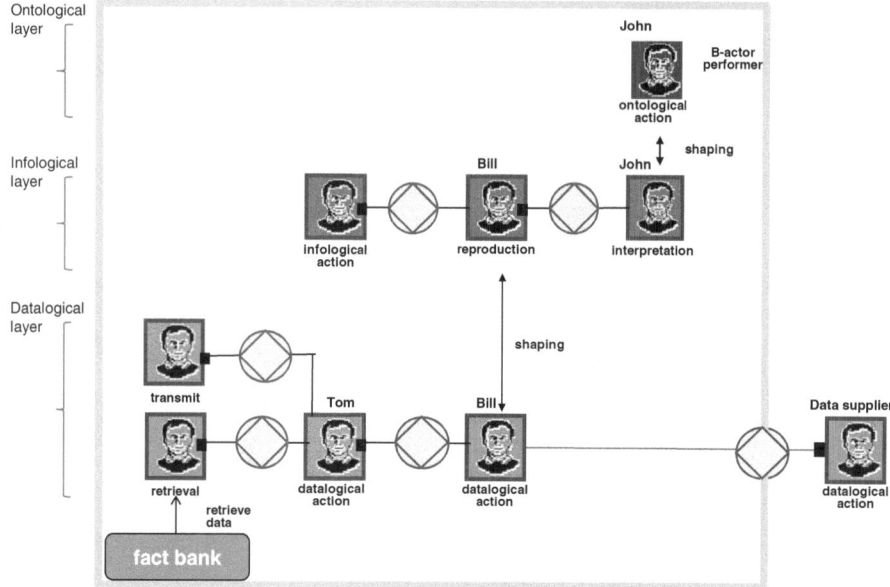

Fig. 9. Asking for internal and external data

delivered by external I-actors. Figure 8 exhibits this situation. John requests both Bill and the info supplier who is drawn outside the organization boundary for information he needs. Both actors deliver their information to John. John joins together both information components and interprets it subsequently. The details about the possible consulted fact banks or about the employment of I-actors outside the organization boundary are of no importance for John. They are outside and for that reason out of scope.

It is important to understand that the infological transaction with the external I-actor regards the operational delivery of the information to John. It does not regard the agreement about the service levels and other conditions which are agreed upon between both organizations. Such an agreement is always the result of an ontological transaction between a B-actor within the current B-organization and a B-actor within the B-organization of the information supplier. The P-fact defined by the performing actor within the B-organization of the information supplier is stored in a fact bank and can be reproduced by John during his infological transaction with the external I-actor. The acceptance by John of the information product offered by the external I-actor takes only place if the delivery occurs according to the agreed conditions. An example of such an information supplier is a financial institute which offers information about the financial stability of a new customer based on a subscription form.

Besides the cooperation between two or more organizations on I-organization level, it is also possible to cooperate with other organizations on D-organization level. This type of cooperation concerns on the level of data exchange. Suppose that an I-actor within the organization wants to reproduce one or more facts, but that the fact data that corresponds with these facts is not obtainable within the organization boundary. Then, this data has been created in another organization and therefore the concerning fact bank is managed by the other organization. Figure 9 exhibits the datalogical transaction between D-actor Bill and the external data supplier. This datalogical transaction regards the operational delivery of data from the external D-actor to Bill. Similar to the situation that is exhibited in figure 8, it does not regard the agreement about the service levels and other conditions which are agreed upon between both organizations. Such an agreement is also the result of an ontological transaction between a B-actor within the current B-organization and a B-actor within the B-organization of the data supplier. The data that corresponds with the P-fact defined by the performing actor within the B-organization of the information supplier is stored in a fact bank and can be retrieved by Bill during his datalogical transaction with the external D-actor. The acceptance by Bill of the data product offered by the external D-actor takes only place if the delivery occurs according to the dataset that corresponds with the agreed conditions. An example of such an external organization that retrieves and records and transmits data is an external data center.

4   Conclusions and Further Research

This paper contributes to more clearness concerning the interactions between the B/I/D-organizations of an enterprise. A B-actor is able to shape into an I-actor and an I-actor is able to shape into a D-actor by the cohesive unification of human being. A B-actor shapes into an I-actor for initiating information requests and interpreting received information products as well as for formulating C/P facts and initiating request for storing these facts. An I-actor shapes into a D-actor for reproducing and remembering facts. A clear understanding about the interactions within and between the layers is important for further research on the determination of both the functional requirements and the constructional requirements of the I-organization based on the construction model of the B-organization. Those requirements are indispensable in order to design and engineer the construction model of the I-organization ultimately.

References

1. Mulder, J.B.F.: Rapid Enterprise Design, Technical University Delft: Delft (2006)
2. Austin, J.L.: How to do things with words. Harvard University Press, Cambridge (1962)
3. Searle, J.S.: Speech Acts, an Essay in the Philosophy of Language. Cambridge University Press, Cambridge (1969)
4. Liu, K.: Semiotics in Information Systems Engineering. Cambridge University Press, Cambridge (2000)
5. Stamper, R.: Information in Business and Administrative Systems. Wiley, New York (1973)
6. Bunge, M.A.: Treatise on Basic Philosophy. A World of Systems, vol. 4. D. Reidel Publishing Company, Dordrecht (1979)
7. Dietz, J.L.G.: Enterprise Ontology – theory and methodology. Springer, Heidelberg (2006)

8. Dietz, J.L.G.: Architecture, building strategy into design, NAF working group Extensible Architecture Framework, xAF (2008)
9. Dietz, J.L.G.: The deep structure of business processes. Communications of the ACM 49(5), 59–64 (2006)
10. Dietz, J.L.G., Barjis, J.A.: Petri net expressions of DEMO process models as a rigid foundation for requirement engineering. In: 2nd International Conference on Enterprise Information Systems. Escola Superior de Tecnologia do Instituto Politécnico, Setúbal (2000)

Author Index